# YHSVH

by

R.L. Worthy

**KornerStone Books**
6947 Coal Creek Pkwy
Suite 206
Newcastle, WA 98059
Ksbooks@execs.com

Published by

## KornerStone Books
**6947 Coal Creek Pkwy**
**Suite 206**
**Newcastle, WA 98059**
**Ksbooks@execs.com**

Copyright ©2008  R. L. Worthy

All Rights Reserved.  No part of this publication may be reproduced, stored in a retrieval system or transmitted in any form or by any means electronic, mechanical, photocopying, recording or otherwise, without the prior written permission of the publisher.

Editor:  Ethel Williams Thompson

Design and Layout:  KornerStone Books

Images courtesy of The Hall of Records -KornerStone©

Printed in the United States of America

*The First Edition*

ISBN: 978-0-9727627-7-9

*Although the learned today believe the keys of knowledge should be hidden—the great peoples of the forefront believed that the gates of knowledge should never be closed—*

*Thus, the Call . . .*

# Introduction

Not one word of this effort is meant to deceive or mislead anyone in the slightest way whatsoever. This is a book about the empowerment of your life through energy—*physical* and *spiritual*. Despite the popular cliché, your body is more than a chance mixture of elements worth 98 cents. Electromagnetic energy is radiating within every cell of your body. Therefore, the quality of your health and experience in this dimension are contingent upon the energies you command—or fail to command—in your life!

Indeed, energy waves have been used to eliminate the cause of many human ailments. In actual point of fact, the energy to improve your health and spirituality are easily obtained—*it is simply up to you to understand this and use it!* For example, did you know that energy has been used to immobilize HIV in humans? *Well, it has!* **Let me repeat that: <u>A way to stop HIV from harming the human body has been discovered</u>**! As for why most people are not aware of it—never forget, there is a tremendous difference between a scientific researcher—and the pharmaceutical industry . . .

Further, *because real life is not a fractional proposition*, YHSVH delves into the energy of words, sound, resonance, and spirituality! No less important, many truths are imparted through this work's maxims, historical art, and endnotes for those of you who value wisdom . . .

*So, if you are ready to put down fake solutions to fake problems whose only purpose is to preserve a fake world—*

# CONTENTS

| | |
|---|---|
| **Dedication** | iii |
| **Introduction** | iv |
| **Reader's Guide** | viii |
| **Words** | 1 |
| **What's in a Name?** | 10 |
| **The Power of Sound** | 25 |
| **Can Energy Waves Heal?** | 34 |
| **YHSVH** | 45 |
| **Babel** | 68 |
| **Under the Sun** | 90 |
| **Epilogue** | 147 |
| **Endnotes** | 153 |
| **Bibliography** | 203 |
| **Index** | 223 |
| **Order Forms** | 233 |

# Reader's Guide to the Work

In the effort to create a no frills straightforward document, I have been forced to employ the use of two time, and space, saving reference symbols: Roman numerals and the word net in parentheses.

A Roman numeral in bold type (e.g.,**I, II, III**) at the end of a sentence indicates an endnote reference citation and that more information about the subject is available. The citation can be found in the corresponding chapter endnote located at the end of the book.

The word net in parentheses (net) at the end of a citation has been used to indicate that the reference can be found on the Internet. The precise url (website address) and date of viewing are supplied in the bibliography.

# Words

## Yeshua

What are words? Words are sounds that convey the ideas, desires and/or intentions of their speaker. To address the earliest non-verbal use of words, they were not expressed by an alphabet (like today) but through symbols (pictures) called *glyphs*. This ancient image/idea association is not very surprising; *indeed, we've all been told that a picture is worth a thousand words.*

Scholars believe that glyph writing was first undertaken in the Nile Valley c. 3500 BC. This unique style of writing was known to the ancient Egyptians as the *Sacred Carvings*. Today it is commonly known as *Hieroglyphic* writing. The ancient Sumerians of Southern Mesopotamia would also develop a writing script c. 3000 BC. Their style of picture (pictograph) writing is known as *Cuneiform*, which means "Wedge" as it was created (or carved) with sharp tools.[1]

---

[1] Budge, E.A. <u>An Egyptian Hieroglyphic Dictionary</u> & Davies, W., <u>Egyptian Hieroglyphs</u> p. 6 & Brander, B., <u>The River Nile</u> p. 157 & Murphy, E.,

# Words

**Egyptian Hieroglyphs c. 2500 BC**

**Sumerian Cuneiform c. 2100 BC**

<u>Diodorus on Egypt</u> p. 104 & Budge, E.A. <u>Egyptian Language</u> pp. 1 - 2 & Arnett, W., <u>The Pre-Dynastic Origin of Egyptian Hieroglyphs</u> & Hieroglyphics, <u>Compton's Encyclopedia</u> Vol. X, p. 150 & Mertz, B., <u>Red Land, Black Land</u> p. 125 & Kramer, S., <u>From the Tablets of Sumer</u> pp. 3, 278 – 279 & Singer, C., <u>A History of Technology</u> Vol. I & Oates, J., <u>Babylon</u> p. 15

## YHSVH

Humans are unique in that words control our lives like no other species on this planet. For the vast majority of us, the sounds we call words influence everything we do, and think, on a daily basis. In actual point of fact, the sounds we call words can carry the power of life and death!

Anthropologists have done exhaustive research in the area of speech and primates. The higher apes are thought to be the closest animals to human beings. While scientists have concluded that apes in captivity can be taught to understand many human words (some of them even learning to use sign language) the ability to actually create the sounds that formulate words is lost upon their biology.

Briefly here, primates are divided into three classifications. The lower classification is called Prosimians; it consists of the lemur, loris and some tree shrews. The higher classification has

# Words

**Photo appears courtesy ZOO Liberec**

commonly been called *Anthropoids* and is comprised of monkeys, apes and humans. Then there is a division of the anthropoids that is known as the *Super-family of Hominoids*. This classification consists only of humans and great apes. Of these apes, it is the chimpanzee (previously pictured) that comes closest to human beings genetically.

Incredibly, chimps share more genetic material with humans than they do with gorillas. The difference between the DNA of humans and chimpanzees is less than 2 percent. I would be remiss here not to point out that because human DNA has about 3 million base pairs, this means that this 2 percent amounts to some 45 million variations (or differences) between the species.[2]

---

[2] Monkeys 'grasp basic grammar' BBC Science/News (Net) & An Apparent New Direct Human Ancestor Fossil Science Week Vol. 5, No. 28, July 13, 2001 (Net) & Candille, S., What Makes Us Human? Tech Museum/Understanding Genetics (Net) & Essential Atlas of Physiology p. 39 & Wikipedia: Throat (Net)

# Words

In assessing the variation we find that two biological traits that the apes do not share with humans are in the formation of their vocal chords and larynx. These organs in chimps are not as large or sophisticated.

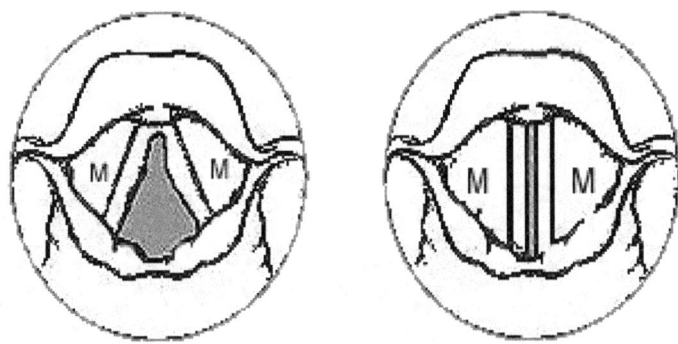

**Left is a top view of the human larynx when open. Right is a top view of the human larynx when closing (M represents the moving mucous membranes).**

Human vocal cords are two mucous membranes which open and close horizontally across the larynx in the upper throat. When humans speak, air is being expelled from the lungs upwards through the vocal chords, larynx and out of the mouth. It is the speed at which your chords vibrate that creates sound as the air rushes through the larynx. Human vocal chords are capable of opening and closing (vibrating) well over 400 times per second! However, anthropologists explain that the smaller and less developed chords, tongue and larynx structure of apes makes this feat impossible for them.[3]

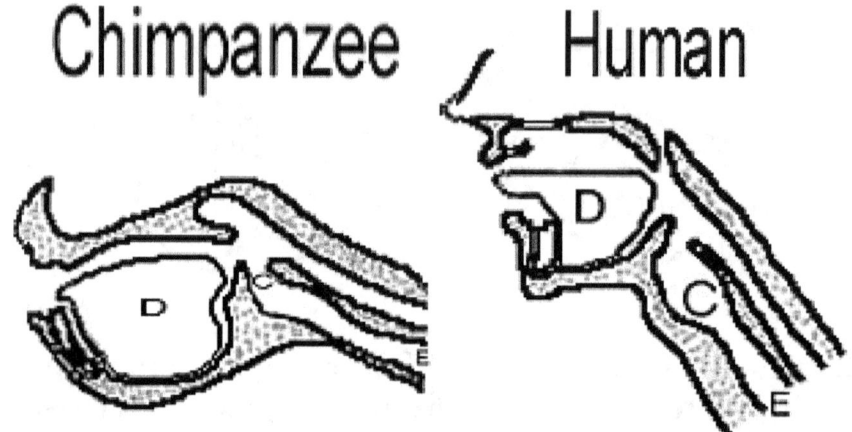

**Cross section view of the chimp and human vocal tracts: C is the position where the vocal chords are seated; D represents the tongue; and E is the trachea (or wind pipe).**

In addition to having a different anatomical structure, humans also possess two important gene formations (or mutations) that the great apes do not have. Both, the FOXP2 and MYH16

---

[3] While the lower primates are not able to form word sounds—the ability to make other calls (and sounds) is crucial to their survival.

## Words

genes in humans are believed to be essential to our ability to comprehend and speak language. However, these gene formations are not present in other primates.[4]

Now, the reason that I took the time to share all of that is to say this. Upon reflection, when we stop to consider that words can literally shape our entire existence—and that of the millions of species on the earth—it is only the human being that can master and produce the dynamic sounds we call words—it is patently clear that words are no small matter! Indeed, allow me to suggest that the only limitations we can ascribe to words are the boundaries created by the small minds who use them . . .

---

[4] Atlas of the Human Body p. 77, 129 & Ghazanfar, A., Primate Audition: Ethnology and Neurobiology pp. 88 - 96 & Morelle, R., Animal world's communication kings BBC (Net) 2007 & Monkeys 'grasp basic grammar' Monkeys can understand the simple rules of grammar but the key element of all human languages is beyond them, a study at Harvard University has shown. BBC Science/News (Net) 2004 & Candille, S., What Makes Us Human? Tech Museum/Understanding Genetics (Net)

# What's in a Name?

# What's in a Name?

Amongst the people of man's earliest civilizations, the first important event in the life of a child was their naming. The reason for this was they believed that existence was not possible without a name. Sayce explains, *"The name, so the Babylonian believed, was the essence of the person or thing to which it was attached; that which had no name did not exist, and its existence commenced only when it received its name."*[1]

**Ancient Egyptian portrait of man with his family**

---

[1] Rosten, L., Joys of Yiddish p. 240 & Woolsey, J., Symbolic Mythology p. 154 & Sayce, A., The Religions of Ancient Egypt and Babylonia pp. 330 - 331

# YHSUH

Erman writes, *"The name possesses something of the nature of the object, and it is owing to this that the sun god . . . is called his own creator because he himself formed his name."*[2] Rosten echoes his sentiment when he tells us that the ancient Hebrews believed that the mystical name of God (YHWH) existed before the sun.

Indeed, similar beliefs were held throughout the Fertile Crescent. After studying many religions in the region, Kramer would describe the matter thus: *"All that the creating deity had to do, according to this doctrine, was to lay his plans, utter the word, and pronounce the name . . ."*[3]

---

[2] Budge, E.A., A Short History of the Egyptian People p. 206 & Sykes, E., Everyman's Dictionary of Non-classical Mythology p. 107 & Muller, M., Mythology of All Races Vol. XII, p. 201 & Erman, A., A Handbook of Egyptian Religion p. 28

[3] Woolsey, J., Symbolic Mythology pp. 154 - 155 & Essenes, Encyclopedia of Religion Vol. V, p. 164 & Kramer, S., From the Tablets of Sumer p. 74 & Genesis 1:28, 2:19-20 & Revelation 2:17

Fertile Crescent is a term that is a designation for the ancient regions of Northeastern Africa, the lands of the Eastern Mediterranean Sea coast (Phoenicia) and Mesopotamia.

## What's in a Name?

Conceptually, this coincides completely with the later Christian teaching found in the *Book of John*, which states:

> *"In the beginning was the Word, and the Word was with God, and the Word was God . . . All things came into being through Him, and apart from Him nothing came into being that has come into being."*[4]

It should also be pointed out that the ancients believed that to name something was to have authority over it. This comes as no surprise; obviously, if the almighty creator names what he creates (from Himself on down) then naming something places the designator in the role of preeminence over that which is named.

This is borne out in Judeo-Christian scripture in the *Book of Genesis*:

---
[4] John 1: 1 - 3

# YHSVH

*"Now the LORD God had formed out of the ground all the beasts of the field and all the birds of the air. He brought them to the man to see what he would name them; and whatever the man called each living creature, that was its name. So the man gave names to all the livestock, the birds of the air and all the beasts of the field."*[5]

Despite creating all life, the fact that the Creator allowed Adam to name all of these creatures is proof that He intended mankind to have authority over this realm of creation. And that control began with the sounds we define as names.[6]

---

[5] Woolsey, J., Symbolic Mythology pp. 154 - 155 & Essenes, Encyclopedia of Religion Vol. V, p. 164 & Kramer, S., From the Tablets of Sumer p. 74 & Genesis 1:28, 2:15 - 20
Parenthetically, we find that this ancient belief has not been wasted on the colonizers of latter times as shortly after they arrive in someone else's homeland—they soon rename it.
[6] Genesis 2: 7
The name *Adam* means "Ground" or "Earth." This is fitting as in the creation story the first man was formed from the earth.

## What's in a Name?

In light of these facts, is it any wonder that the ancients should give the utmost consideration to the names they gave their children?  Likewise, we find the ancients trying to carry themselves in ways that would not tarnish their character, or *"Good Name."*  Indeed, on top of the many axioms that the ancient Egyptians had for living a good life—we find the common salutation, *"May my name prosper!"*[7]

We even find this standard being evidenced in the *Book of Psalms*, when it is explained: *"He leads us in the path of righteousness for His name's sake."*  Actually, the Judeo-Christian scripture makes reference to the sanctity of the name more than two dozen times.[8]

---

[7] Rosten, L., Joys of Yiddish p. 240 & Woolsey, J., Symbolic Mythology p. 154 & Sayce, A., The Religions of Ancient Egypt and Babylonia pp. 330 - 331 & Bunson, M., The Encyclopedia of Ancient Egypt p. 157 & Budge, E.A., The Dwellers on the Nile p. 36
[8] Psalms 23:3

## YHSVH

Let's take a moment to look at the kind of names that were bestowed upon the great people of the societies of the First World:

**Queen Hatsheput c. 1480 BC**

This is a limestone bust of Egypt's Queen Hatsheput. Ruling for more than two decades, she is considered to be the first great female

# What's in a Name?

monarch in history! The name Hatsheput means, *"Above the Measure of the Noblest Women."*9

**Pharaoh Akhenaten c. 1400 BC**

---

9 Kaster, J., Wings of the Falcon: Life and Thought of Ancient Egypt p. 117 & Montet, P., Lives of the Pharaohs p. 81 & Budge, E.A., A Short History of the Egyptian People p. 74 & ben-Jochannan, Y., Black Man of the Nile and His Family p. 176 & Budge, E.A., Egypt pp. 118 - 121 & Trigger, B., Kemp, B., O' Connor, D., & Lloyd, A., Ancient Egypt: A Social History p. 184 & Fairservis, W., The Ancient Kingdoms of the Nile: And the Doomed Monuments of Nubia p. 167

## YHSVH

The stone statue on the previous page is of the famous Egyptian Pharaoh Akhenaten. Though extremely controversial in his day, many latter day historians have come to hold him in high regard. There are two principal reasons for this: first, that he did not use the Egyptian military (strongest in the world at that time) against other nations; and second, he was the first Egyptian king to champion monotheism. The name Akhenaten means, *"Glorious for Aten."*[10]

Looking to the east, while the laws of Lipit-Ishtar, Ur-Nammu, and Eshnunna were fine achievements—it is the famed Mesopotamian

---

[10] Montet, P., Lives of the Pharaohs pp. 138 - 139, 141 & Budge, E.A., A History of Egypt Vol. IV, pp. 118 -119, 135 - 140 & Sykes, E., Everyman's Dictionary of Non-classical Mythology p. 21 & Akhenaten, The Encyclopedia Britannica Vol. I, p. 189 & Hobson, C., The World of the Pharaohs p. 107 & Gardiner, A., Egypt of the Pharaohs pp. 230 - 232 & Budge, E.A., A Short History of the Egyptian People pp. 94 - 95
Akhenaten is also spelled, Ikhnaten. This king is regarded as history's first pacifist ruler. Monotheism is the worship of one God.

## What's in a Name?

**King Hammurabi c. 1700 BC**

Law Code of Hammurabi that is most heralded today! Creating about 300 clauses, the legal tenets of Hammurabi's kingdom were completely established in it: 5 sections addressing the administration of justice; 20 sections outlining property offenses; 50 sections on land tenure; about 40 sections on trade and transactions; 68 sections on family matters (including marriage, divorce and adoption); 20 sections on assault; 16 sections on proper payment for services rendered; 5 sections on the treatment of slaves, and so forth. This king may also have been the first ruler to employ the use of the seven-day calendar week. Fittingly, the name Hammurabi means, *"Uncle is the Healer."*[11]

Another dynamic figure of the forefront was Rameses II. One of the most important names given to this pharaoh at his coronation was Usimare-Setpenre, which is translated as *"Strong is the Truth of God—Elect of God."* This name was placed in the first of the king's two

---

[11] Mieroop, M., King Hammurabi of Babylon & Richardson, M., Hammurabi's Laws: Text, Translation and Glossary & Oates, J., Babylon & Kramer, S., From the Tablets of Sumer & Hammurabi, Behind the Name: the etymology and history of first names (Net)
It has been advanced that the noses of many stone artifacts have been deliberately damaged to obscure the racial identity of the person that's portrayed.

## What's in a Name?

cartouches.[12] Mastering his foreign and domestic obstacles before the 30th year of his reign, Rameses would experience decades of pageantry, pleasure, prosperity and peace![13] Whether in building, luxurious living, or even the blessing of royal progeny—Rameses II's deeds were unmatched. No less significant, directing a great army and navy, his royal decrees would be dutifully carried out from the

---

[12] De Lubicz, R., <u>Sacred Science: The King of Pharaonic Theocracy</u> p. 167 & Baines, J., & Malek, J., <u>Atlas of Ancient Egypt</u> pp. 36 - 37 & Montet, P., <u>Lives of the Pharaohs</u> p. 95, 134, 204 & Trigger, B., Kemp, B., O'Connor, D., & Lloyd, A., <u>Ancient Egypt: A Social History</u> pp. 56 - 57 & Budge, E.A., <u>A Short History of the Egyptian People</u> p. 27, 193 & Quirke, S., <u>Who were the Pharaohs</u> p. 10 & Gardiner, A., <u>Egypt of the Pharaohs</u> p. 14, 52 & Budge, E.A., <u>The Mummy</u> & Budge, E.A., <u>The Dwellers on the Nile</u> p. 87 & Mertz, B., <u>Temples, Tombs and Hieroglyphs</u> pp. 23 - 24 & Budge, E.A., <u>A History of Egypt</u> Vol. II, p. 19 & Bunson, M., <u>The Encyclopedia of Ancient Egypt</u> p. 86, 215, 230 & ben-Jochannan, Y., <u>Black Man of the Nile and His Family</u> pp. 159 - 157
The cartouche was the circular hieroglyphic emblem that was used to designate each king's name. It was a rope knotted at two ends since double-knotted ropes were seen as symbols of great power. The circular pattern was chosen for the cartouche because it represented eternity.
[13] Montet, P., <u>Lives of the Pharaohs</u> p. 194

## 𝔜𝔥𝔖𝔲𝔥

Nile Valley's 5th cataract to the Northern Levant (a distance of some 6,000 miles)![14]

**Rameses II c. 1200 BC**

---

[14] Trigger, B., Kemp, B., O' Connor, D., & Lloyd, A., <u>Ancient Egypt: A Social History</u> p. 184 & Budge, E.A., <u>A History of Egypt</u> Vol. V, p. 71

The ancient Northern Levant equates to the modern day region of Syria. Coming to the throne in his early twenties and ruling Egypt for somewhere between 65 and 67 years, Rameses II died very near the age of one hundred!

## What's in a Name?

There isn't any question that the people of many lands have recognized the significance of the sounds we call names. For instance, the ancient Chinese believed that wisdom began with calling things by their right names.[15] Further here, naming ceremonies have been conducted throughout Africa for millenniums.

Many of you will recall the naming ritual re-created in *Roots*, where on the eighth day of his life, Kunta Kinte's father Omoro held him up under the night sky, pronounced Kunta's full name and declared—*"Fend killing dorong leh warrata ks itch tee"* which means, *"Behold the only thing greater than yourself!"*[16]

Actually, the christening of children by some in the West can be seen as a ceremonial naming.

---

[15] Seldes, G., The Great Quotations p. 678
[16] Haley, A., Roots pp. 12 - 13 & Roots (VHS) 1981
You will also recall that Kunte would be ostracized and beaten for refusing to accept the slave name of Toby.

## YHSUH

We even find the custom being applied to inanimate objects like ships, airplanes and buildings.

In closing, the case is easily made that few things are more pivotal than names. Upon reflection, I am reminded that the Egyptians believed that any soul that did not have a name was lost because it could not receive eternal life: *to be judged, one first had to have a name!*[17]  Of course, in these times the mere mention of the correct name can open doors in the human experience. On the other hand, apprehension and/or misgivings are just as certain to follow the expression of the wrong one. Lastly, names distinguish us from our cradles to our graves and beyond. So, with this all shared, should you still need to ask me—*What's in a Name?* Without hesitation I must insist—*Everything!*

---

[17] Budge, E.A., The Dwellers on the Nile p. 36

# The Power of Sound

# JESUS

I believe that we have established that words and names are important in time and space. That achieved, the next question becomes—*What are words made of?* The answer is, *Words are made up of sounds*. Ergo, what we have actually established is that *sounds are important in time and space*.

Now—for so much as an inquiring mind might just want to know—W*hat is sound?* Sound is an invisible form of energy that travels through the air like a wave until it hits the ear. Sound waves move through the air around 750 miles per hour. This is why there is no perceivable time delay between the mouth's moment and the ear's hearing in normal conversation.[1] Once the sound wave comes in contact with the ear, the vibrations created upon the eardrum send

---

[1] NASA: Sound - Speed of (Net) & Physics Classroom tutorial: The Nature of a Wave (Net)
The speed of sound is affected by altitude and temperature. A wave is a disturbance that moves through a medium and transports energy from one point to another.

signals to the brain. Those vibratory signals are interpreted as distinct sounds: a car door closing, hands clapping, a woman's voice, a child's laughter, a cat purring, etc., etc.

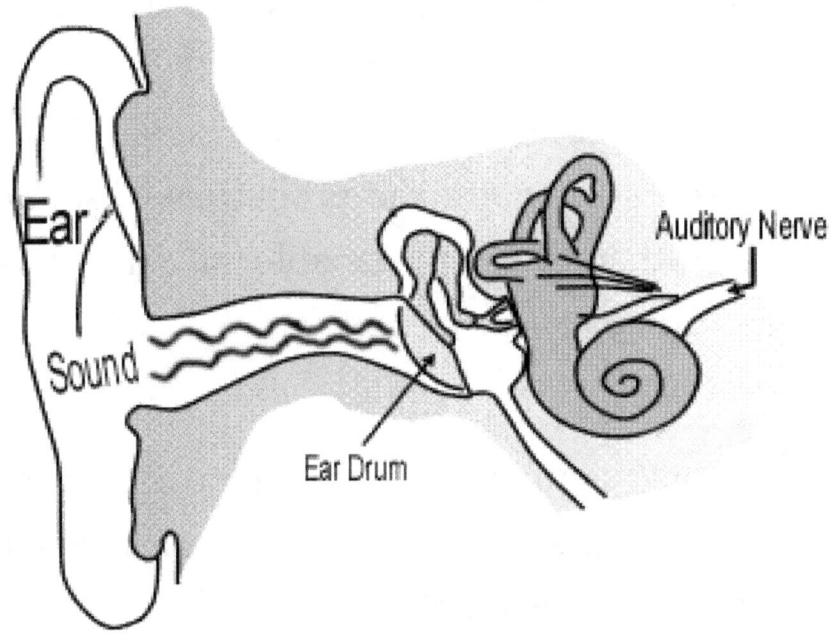

Because most of us take hearing for granted, no doubt some of you are thinking, *So sound energy makes noise—what's the big deal?* Well, let's see if you can hear this. Remember those Memorex commercials with Ella Fitzgerald back in the 70s? All right, maybe some of you weren't

# YESUR

born in the 70s, but how about those movies where an overweight singer's voice is so powerful that once they hit, *and hold,* a certain note—a crystal glass across the room breaks? Scenes like this are produced in American comedies from time to time; however, let me assure you that fat lady or no—the subject is no joke! You see, one of the properties of sound waves is that they are capable of shattering things! *Can you hear me now?*

The photo sequence on the opposite page records what happens to a crystal glass when subjected to resonant sound wave energy. Scientifically speaking, what's occurring is that the resonance point of the matter (crystal glass) is being stretched beyond its limit by the pitch (tone) and amplification (force) of the sound wave energy. In layman's terms, the glass has a natural resonance point (or frequency) at which it vibrates.

# The Power of Sound

[2] <u>Essential Atlas of Physiology</u>  pp. 82 – 83
This entire sequence took about 20-seconds.  The images are provided courtesy of Acoustics, Audio and Video Group, University of Salford - www.acoustics.salford.ac.uk

# JESUS

When the proper sound wave energy is directed at the glass and amplified beyond its safe resonance (vibratory) point for a sufficient period of time—the vibration ultimately causes the glass to shake apart, or shatter. However, a key point here is that just creating a loud noise will not shatter glass. It is only sound at the proper pitch and amplification that will produce the wave energy needed to move the matter to the resonance point and cause it to disintegrate!

Assuming acoustics is new to you, allow me to take a moment to say a word about frequency and resonance. The higher the frequency in Hz (Hertzs) the higher the pitch. The lower the frequency in Hz, the lower the pitch. For the sake of our discussion, think of one Hz like a second. If I say that the frequency of a certain note is 300 Hz, I am telling you that the sound wave has 300 cycles per Hz. If say another note has a frequency of 600 Hz, I have told you that it has 600 cycles per Hz. I have also told you that the latter has a higher pitch than the former.

**These symbols represent musical notes. The second note from the right is the lowest pitched note of the four so its wave cycle frequency is also slower in Hz.**

Scientists believe that all matter has a resonance point. In the case of a crystal glass, the amplification of one frequency (pitch or tone) is all that's needed to shatter it. However, in theory—items made up of several elements might well require a combination of amplified frequencies to cause their disintegration.[3]

---

[3] Rumsey, F., & McCormick, T., <u>Sound and Recording: An Introduction</u> Ch. 1 & Clark, J., Yallop, C., & Fletcher, J., <u>An Introduction to Phonetics and Phonology</u> pp. 302 - 304 & Wikipedia: <u>Hertz</u> (Net) & Wikipedia: <u>Frequency & Fundamental Frequency</u> (Net) & Physlink.com: <u>What is the physics involved with breaking glass with your voice</u>? (net)

## YESHUA

While most of us have seen demonstrations of a loud sound shattering something, few of us think of sound as a profound force that is capable of destroying matter. Yet, sound is a mechanical energy wave that is absolutely capable of that! Thus, anyone who can command sound energy is capable of that! *Do you hear me now?*

I realize that some of the spiritual people in the house may be thinking that this is a little too scientific, *or far out*, to spend much time contemplating; yet, what do your scriptures tell you? In case you need a refresher they explain that sound waves were instrumental in the collapse of the great walls of Jericho! Not only that, we also find: *"Out of his mouth came a sharp double-edged sword."*[4]

The significance of the latter to our discussion is that the Bible clearly depicts the double-edged

---

[4] Joshua 6: 4 – 20 & Revelations 1:16

## The Power of Sound

sword as a deadly weapon. Also, we know that the mouth is the organ where vibrations (created in the vocal cords) leave us as sound wave energy. Hence, it is hardly blasphemous to suggest that this same passage could be translated—*Out of his mouth came a sound wave that's so powerful it can destroy!*

*Lastly, one final point needs to be made here about sound wave energy. Those who wish to prey on you understand that humans are a sound controlled species; ergo, it is their aim to keep you phonologically and intellectually ignorant. This is so that (as your supposed masters) they can deceive you into either behaving like a mindless animal that's led to and fro—or, so that you will follow them (lockstep) into total ruin.* <u>*The only way to triumph over them is to vigilantly test the sounds and spirits that you encounter—embracing the true and rejecting the false—come what may*</u> *. . .*

# Can Energy Waves Heal?

## Can Energy Waves Heal?

Having established that sound is a wave of energy that is capable of disrupting matter, medical researchers have asked the question—*If sound waves can be used to destroy, can they be used to heal?* The answer to the question is a resounding—*Yes!*

Remember going from fine to shattered crystal with less than 20 seconds of sound wave energy?

---

[1] Images courtesy of Acoustics, Audio and Video Group, University of Salford - www.acoustics.salford.ac.uk

Well, extracorporeal shock wave lithotripsy (ESWL) has been used for some time now to break up stones in the kidneys, bladder, gallbladder, ureters and liver. An extracorporeal shock wave is (you guessed it) *a sound wave.* The only caveat here is that ESWL is performed with ultrasound. Ultrasound is a category of sound that's above the range of human hearing. The range of hearing for humans is from about 20 to 20,000 kilohertz. Ultrasound devices emit sound waves that operate in ranges from 2 to 18 megahertz.[2] However, the principle is exactly the same as what's been previously explained, once a stone's resonance frequency is reached and maintained—*it shatters!*

Another therapeutic application for sound is meditative. This practice has been utilized for thousands of years. In fact, the learned peoples

---

[2] Webster's Medical Dictionary p. 414 & Watson, T., Therapeutic Ultrasound (Net) & Rapacholi, M., Essentials of Medical Ultrasound: A Practical Introduction to the Principles, Techniques and Biomedical Applications pp. 141, 160 – 162 & Loyola Medicine: Kidney Stones (Net)

## Can Energy Waves Heal?

of the East have long considered meditation to be crucial to their wellbeing. What's more, we even find Judeo-Christian scripture championing the practice of meditating on the Creator's name and deeds!

Many in the West see meditation as a waste of time. You know, some Eastern person sitting in the dark making funny sounds—or merely laying down catnapping the day away; however, this is not what meditation is. In actual point of fact, the root word for meditation is *Med-Eri*, which means "to Remedy" or "to Heal."

We have found that sound is an invisible wave of energy. Further, we know that once that wave contacts an object the result is vibration. The ancients understood that the sound waves humans create do not only travel outward—*but inwards as well*. Hence, truly meditating (in the ancient sense) is to create wave energy that

# 𝔜𝔥𝔖𝔳𝔥

produces vibrations within our cells, which promote comfort and/or healing! Permit me to share a statement here by Jonathan Goldman:

> *"In various specialized workshops, I had groups chant 'YHSVH' . . . In addition to its ability to produce divine frequencies . . . it also seemed to specifically stimulate the thymus gland's related energy center and anchor . . . This energy is able to do a vibrational recalibration of our cellular structure and DNA."*[3]

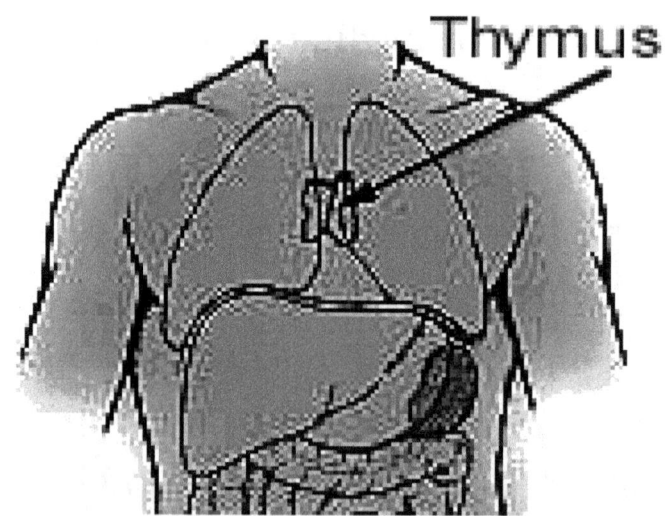

**Physicians explain that the thymus gland is vital to the immune system of human beings.**

---

[3] Goldman, J., Holy Harmony (CD) & Webster's Medical Dictionary p. 400

## Can Energy Waves Heal?

Taking energy waves even further, researchers who are combating the deadly new rash of retroviruses are claiming that electromagnetic (or radio) waves might well be a key to the safe elimination of these harmful invaders from the body! The principle difference between electromagnetic wave energy and the sound wave energy we've previously discussed is that electromagnetic energy waves are produced by something that is electronically charged. Electromagnetic waves can also travel through a vacuum while sound waves will not pass through a vacuum.[4]

The most serious of the newest retroviruses is HIV (Human Immunodeficiency Virus). People who contract this virus commonly develop Aids (Acquired Immune Deficiency Syndrome). Unfortunately, HIV has led to the deaths of

---

[4] Prentice, W., Therapeutic Modalities in Rehabilitation pp. 3 - 12

millions of people worldwide; <u>however, a way to immobilize HIV has been discovered!</u>

Back in the 1980s, Dr. Robert Strecker explained that several decades before, an inventor named Royal Raymond Rife would make the declaration that he could destroy living viruses by subjecting them to radio waves. Rife was also to explain that this method did not harm the human host. Of course, many scientists scoff at the notion of an invisible energy wave being used to shatter a virus. Yet, *we have only just seen ESWL shatter stones and sound waves shatter crystal glass.* All that was needed was the proper frequency and amplification—and Strecker believed that this principle would work against HIV.

**Example of a cross section view of HIV
Note the outer wall's clear crystalline structure**

## Can Energy Waves Heal?

Despite the fact that HIV is a biological infectious agent that has the ability to replicate—its anatomical structure is *crystalline*. Now what's interesting about that finding is that acoustical engineers have found that glass made out of crystal will shatter at frequencies that other types of glass will not. So, the question becomes—*Are organisms with a crystalline formation more susceptible to an energy wave than organisms that do not have this structure?* According to Royal Raymond Rife, Robert Strecker (and other researchers) the answer is—*Yes!*[5]

---

[5] Lynes, B., The Cancer Cure That Worked pp. 106 – 107 & The Strecker Memorandum (VHS) 1988 & Tong, L., Pav, S., Pargellis, C., Lamarre, F., & Anderson, P., Crystal structure of human immunodeficiency virus (HIV) type 2 protease in complex with a reduced amide inhibitor and comparison with HIV-1 protease structures. Proceedings of the National Academy of Sciences Vol. 90(18) Sep. 15, 1993 & Stanfield, R., Gorny, M., Pazner, S., & Wilson, I., Crystal Structures of Human Immunodeficiency Virus Type 1 (HIV-1) Neutralizing Antibody 2219 in Complex with Three Different V3 Peptides Reveal a New Binding Mode for HIV-1 Cross-Reactivity American Society for Microbiology Vol. 80(12) Jun. 2006 & Horowitz, L., DNA: Pirates of the Sacred Spiral (DVD) 2005
I would be remiss not to tell you that Robert Strecker had his detractors. Publicly stating that the WHO (World Health Organization) was responsible for introducing HIV into the human population through its vaccination programs is no way to make, or keep, *"friends"* in high places. Yet, the fact that Judas would betray his Master with a kiss was not lost upon Strecker (pp. 154 –155).

But the most crucial finding here is that the outer structure of HIV is indeed crystalline (page 40). Therefore, regardless of the origin of the virus, Strecker's position was, and still is, not without merit. <u>*In actual point of fact, the outer surfaces of scores of dangerous viruses are substantially crystalline; i.e., possessing symmetrical limbs and appendages*</u>.

Remarkably, a few years after the pronouncements of Strecker, two scientists used an energy wave to achieve what he theorized. <u>In 1990, William Lyman and Steven Kaali (researchers at Albert Einstein College of Medicine in New York) found a way to stop HIV!</u> **These researchers immobilized the Human Immunodeficiency Virus by introducing low voltage DC current into human blood infected with HIV**.

Lyman and Kaali found that when they applied DC in the range of 50 - 100 microamperes (uA), the outer crystalline structure of the HIV was damaged to such an extent that it was unable to attack a normal cell. *Once the virus was immobilized, the body's immune system then took over and naturally expelled the HIV from the body.* Further, at <u>50 to 100 microamperes of electric current the body's healthy cells were not adversely affected:</u> ***no side-effects***! While introducing electricity to the circulatory system

# VESVS

is hardly child's play—Kaali and Lyman have conclusively demonstrated that even in the case of HIV—*Energy waves can help heal!*[6]

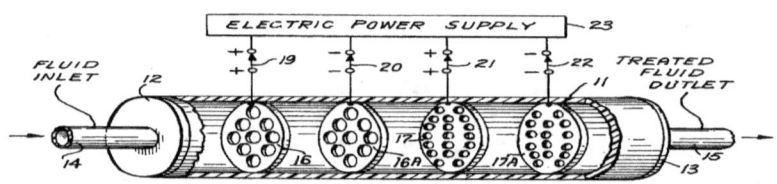

This is Kaali and Schwolsky's patent for an implantable electrifying device that makes HIV inoperable. It was granted by the United States Patent Office in 1992. As noted by Adachi—<u>Obtaining a U.S. Patent is no walk in the park. If this device did not work precisely as claimed—a patent would never have been granted!</u> Today, there are other ways to electrify blood; however, we all are indebted to Kaali, Lyman, and Schwolsky.

---

[6] <u>Shocking Treatment Proposed for Aids. (Zapping the Aids Virus with low voltage electric current)</u>. Science News Vol. 139, No. 13 Mar. 30, 1991 p. 207 & Adachi, K., rexresearch.com: <u>The Story of Blood Electrification</u> (Net)
I also want to acknowledge <u>Dr. Robert C. Beck</u> for his later contributions in this area (*see* Under the Sun).

# YHSVH

### YHSVH

*There's a saying that's been popularized in these times by the motivational speaker Les Brown which goes, "If you want to keep on getting what you've been getting—keep on doing what you've been doing." But if where you are isn't where you want to be—keep on reading . . .*

# YHSVH

Let me applaud your decision to turn the page because you had the courage and faith to believe that <u>no child of the Creator need ever fear the truth</u>! Ergo, your victory is at hand . . .

We have seen that sound waves can control our human experience. Further, in the purely physical sense, we have seen that sound wave energy can cause things to disintegrate. We have even found that other forms of wave energy can destroy and heal! Now let me ask you a question—*If the magnitude of wave energy is this profound in the temporal (as in temporary) plane—do you think that it is more, or less, significant in the eternal realm?*

**The Milky Way Galaxy**

# YHSVH

Allow me to suggest that sound is extremely important in the eternal realm! As a matter of fact, what did the Apostle John tell us:

> *"In the beginning was the Word, and the Word was with God, and the Word was God. The same was in the beginning with God. All things were made by him; and without him was not any thing made that was made . . ."*[1]

Now, what have we established about the essence of words? We have established that words are wholly made up of sound. Hence, it is not misleading to interpret John as explaining:

> *In the beginning was the Divine Sound, and the Divine Sound was with God, and the Divine Sound was God. The same was in the beginning with God. All things were*

---

[1] John 1: 1 – 3
Image of the Milky Way Galaxy on the preceding page appears courtesy of the National Space Science Data Center – 89-089A-01A, 89-089A-02H, 89-089A-03D.

# YHSUH

*made by him; and without him was not any thing made that was made . . .*

It is not possible to refute this once reflecting upon the teachings in the *Book of Genesis*:

*"And God said, "Let there be light," and there was light. God saw that the light was good, and He separated the light from the darkness . . ."*[2]

To say is to speak, and the process of speaking words cannot be accomplished without sound. Thus, we must conclude that divine speech is the formation of divine sound into words of power. Once He says, *"Let there be"*—it is! However, before He initiates the divine sound into action—it isn't.

So as not to be accused of turning a deaf ear, or blind eye, to those of you who are atheists—let

---

[2] Genesis 1: 3 - 4

me express this another way. The most popular scientific view of the beginning of creation is that it all began with a "Big Bang!" *Webster's Dictionary* describes a bang as *"a sudden loud noise."* Obviously, a sudden loud noise cannot occur without the presence of sound: *no sound, no bang.*

All right, so you want to know the point of all of this? Here it is: *Whether you are spiritual or not—you must concede that sound has been found to play a vital role on both, the temporal and celestial, planes—**Be or Bang**!*

**The Earth**

That all said, let me ask this—*Do you think that every sound is exactly the same?* The clear answer to the question is a resounding—*No!* We have seen that sound wave energy has the power to disintegrate glass. However, remember, it was not simply a loud noise that shattered the crystal. <u>Nothing happened to it until the proper sound wave pitch and amplification hit the glass</u>.

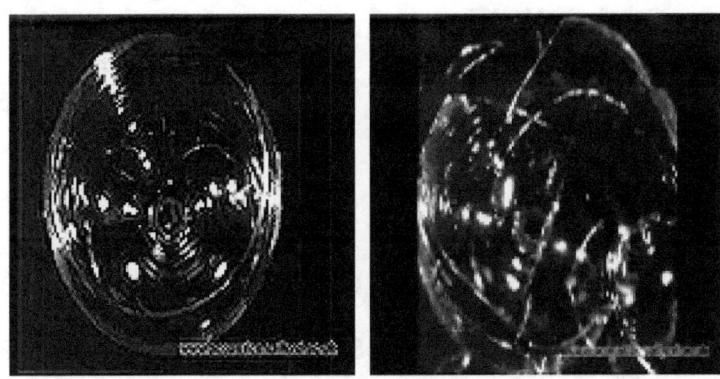

**Two images of a perfectly fine crystal glass being shattered by a sound wave**

Understanding that every sound is obviously not the same, permit me to take a moment to

---

[3] The image on the preceding page appears courtesy of National Space Science Data Center - A17-h-148-22725. The images above appear courtesy of Acoustics, Audio and Video Group, University of Salford - www.acoustics.salford.ac.uk

demonstrate that even the way sound is strung together affects its ultimate effect.  Let's say I told someone to *"Just Be Happy!"* Now suppose that I meant to repeat this remark to someone else but, for some reason, I said *"Happy be Just!"* While the first comment was self-explanatory, the latter is not.  Although I have used the same sounds and even the exact same number of syllables—the mere reversal of the sequence of the sounds has obscured my original intent.

Should you be wondering why I have so painstakingly tried to establish the importance of sound, here is my reply—*Nothing of greater importance has been more misunderstood in these times than the true name of the Anointed One of the ancient Hebrews—YHSVH!* This has resulted in our not utilizing the Creator's gifts to us to their fullest potential—*but Oh, what a difference the day makes!*

## YHSUH

To begin at the beginning, the ancient Hebrews believed that the true name of the Creator of all things was *YHVH* (also written as *YHWH*). The Hebrew pronunciation of this name is Yod-Hey-Vav-Hey. Several translations for the meaning of this name can be found today; e.g., "He who brings into existence whatever exists," "He who brings the Hosts into existence" and "To Breathe is to Exist" are often cited.

**These are the four Hebrew letters that represent the name YHVH. Over time, the name has become known as the *Tetragrammaton*, which simply means "Four Letters" in Greek.**

It has been stated that the Hebrews thought the name of the Creator possessed tremendous power. Because the Hebrews held the meaning and sound of the tetragrammeton in such great reverence—it was hardly ever spoken or written down in ancient times. In fact, many believe that its correct pronunciation was a secret only known to the highest priests.[4] Eventually, some written abbreviations like *Yah* and *Yo* were used as substitutions for the holy name. By the time we get to the Christian era, <u>new designations like *Lord* and *God* had completely replaced the tetragrammeton and its Hebrew abbreviations</u>.

Today, the ancient YHVH (Yod-Hey-Vav-Hey) of the Hebrews has become the *Yahweh* (Yah-Way) of many Western Christians. Incidentally, the name *Jehovah* is the Latin word for YHVH (*IHVH* in the Latin alphabet). Yet, not coming into existence before the past millennium, many linguists are critical of its use.[5]

---

[4] Yahweh, <u>Anchor Bible Dictionary</u> Vol. VI, pp. 1011 – 1012 & Horowitz, L., <u>DNA: Pirates of the Sacred Spiral</u> (DVD) & Yahweh, <u>Encyclopedia Britannica</u> Vol. XII, p. 804 & Yahweh, <u>New Catholic Encyclopedia</u> Vol. XIV, p. 883
Many of the writings of the ancient Hebrews are lacking vowel sounds. Yet, insomuch as most of the vital teaching was transmitted orally—the fact that every word wasn't spelled out would not have been a problem for the learned Hebrews of the time.
[5] Ibid.,

# YHSVH

Moving forward, the Hebrew name of the Anointed One (the Christ) is *YHSVH*. This name means, "YHVH Saves" or "YHVH is Salvation." The scriptures explain that the name of the Anointed One was what made it possible for the apostles to perform miracles. Word: <u>When the Christ's apostles were performing said—Yod-Hey-Shin-Vav-Hey was the powerful declarative sound wave of authority they were proclaiming</u>!

As for why this understanding has been lost on most today, all of the letters of the Hebrew alphabet do not have Greek equivalents. Once departing from the original tongue, the tonal syllabic, and vibratory expression of YHSVH was changed. He who was first called Yod-Hey-Shin-Vav-Hey began to be known as *Iesous*, which is pronounced Ee-a-Sous in Greek. In turn, the Greek Iesous would become the root word for the later Anglo-Saxon name *Jesus*, which is commonly pronounced Gee-zus.

# YHSVH

Yet, there isn't any doubt that the sounds created by the name Jesus (Gee-zus) were not upon the lips of the apostles when they were performing miracles in the name of YHSVH. In light of this undeniable historical fact, is it any wonder that Crowley should make this remark:

> *"Other than the Holy name of God himself, no other name has been more called upon through the centuries than that of the Jewish teacher . . . Equally important, would perhaps be a return to the correct Hebrew pronunciation of the Messiah's true name, as a word of power."*[6]

---

[6] Jesus, <u>Anchor Bible Dictionary</u> Vol. III, p. 773 & Goldman, J., <u>Holy Harmony</u> (CD) & Horowitz, L., <u>DNA: Pirates of the Sacred Spiral</u> (DVD) 2005 & Crowley, B., <u>Words of Power: sacred sounds of East & West</u> p. 209
Incidentally, many believe that *Iesous* originally meant, "Healing Zeus" in Greek. Ancient Hebrew was read from right to left, or the opposite of English. Hence, the small symbol furthest right is the Yod sound in YHSVH (Yod-Hey-Shin-Vav-Hey).

## YHSUH

Now, because much of what we've discussed has been about the fact that every word in creation is not the same—let's take a moment to see what sound wave energy actually looks like! The ensuing pages contain computer representations of sound waves created by a person's voice in a professional sound studio. Using the graph format, the vertical line represents amplitude (A) while the horizontal line represents time (T). The energetic line represents the wave energy of the word. The time measure is variable with each word; however, the graph image clearly reveals the unique nature of the sound wave energy of each word.

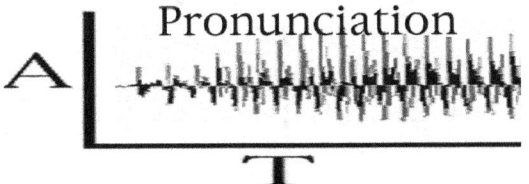

**Example of Sound Wave Graph**

**One quick point—each graph is like a picture. If you take one picture of a baby and another of a man, you can fill up both picture frames—however, the baby is still smaller in stature. The same applies here to mono and multiple syllabic words.**

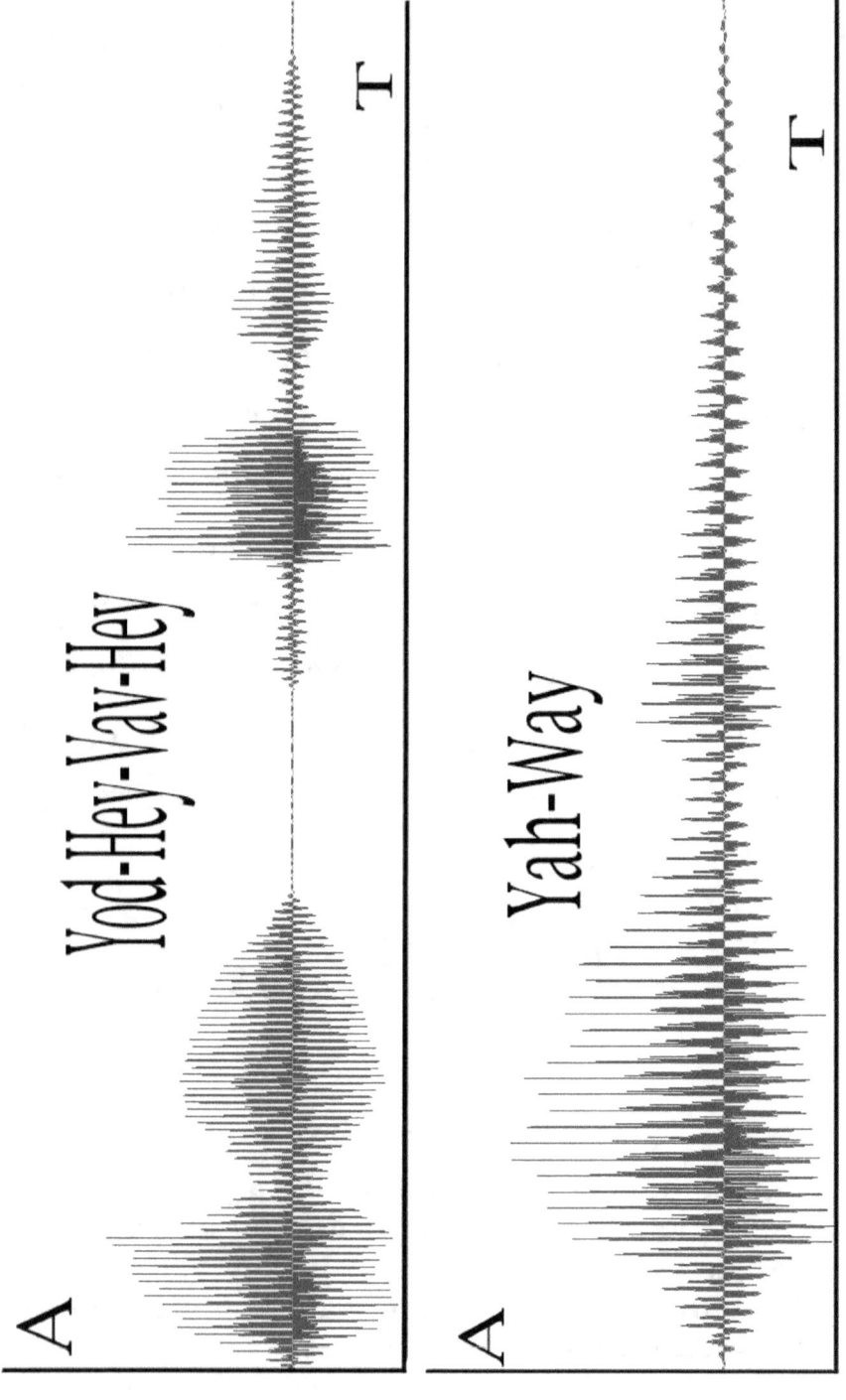

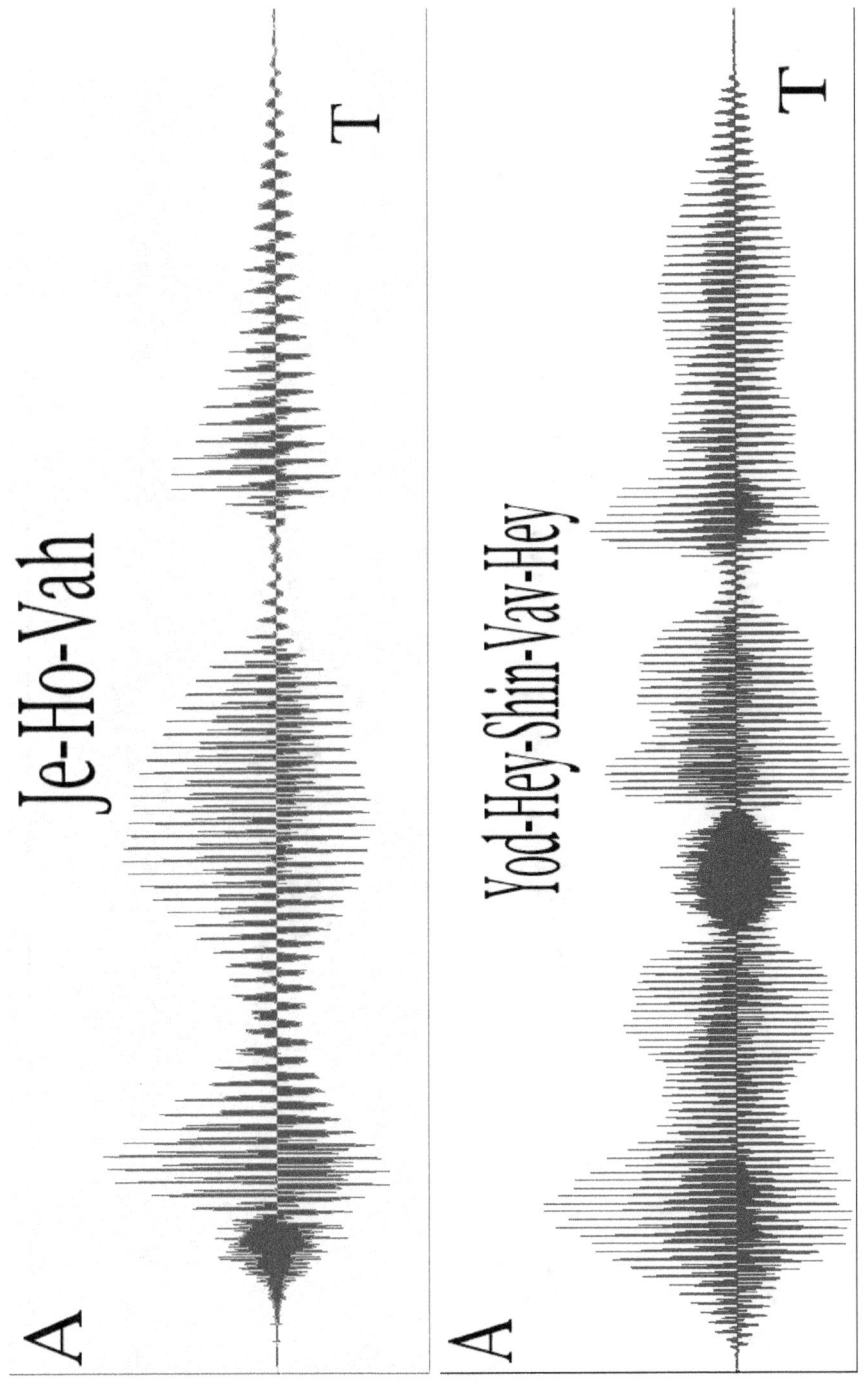

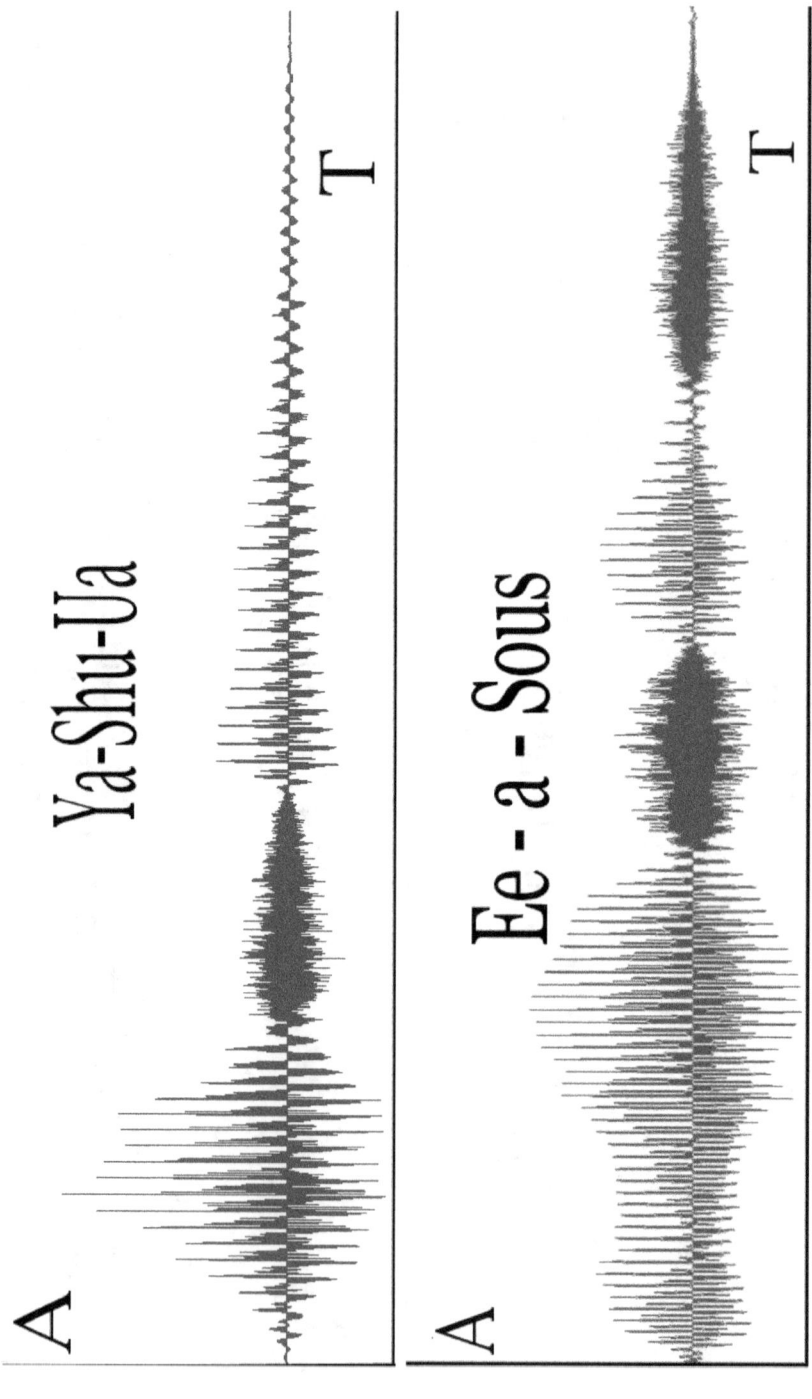

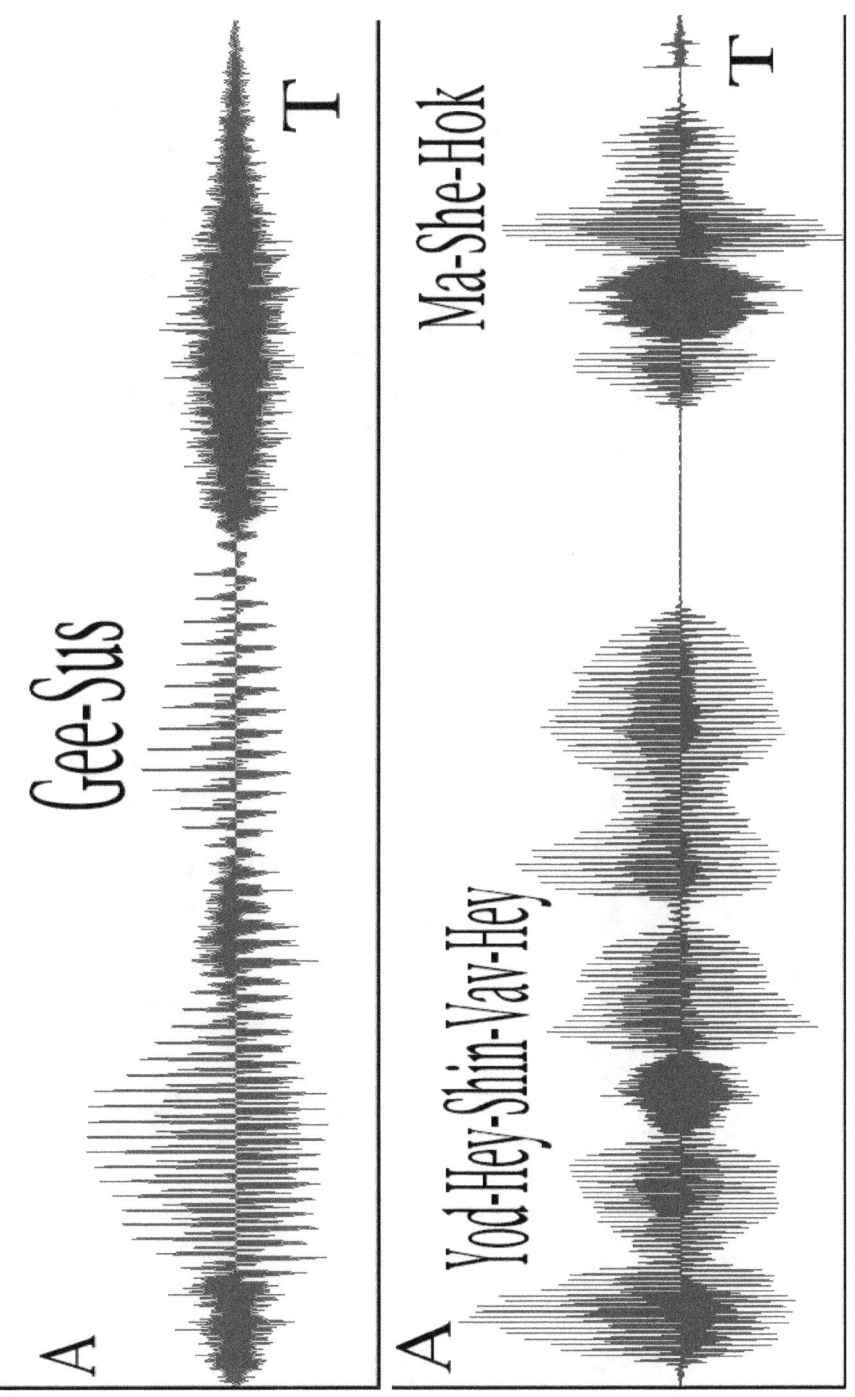

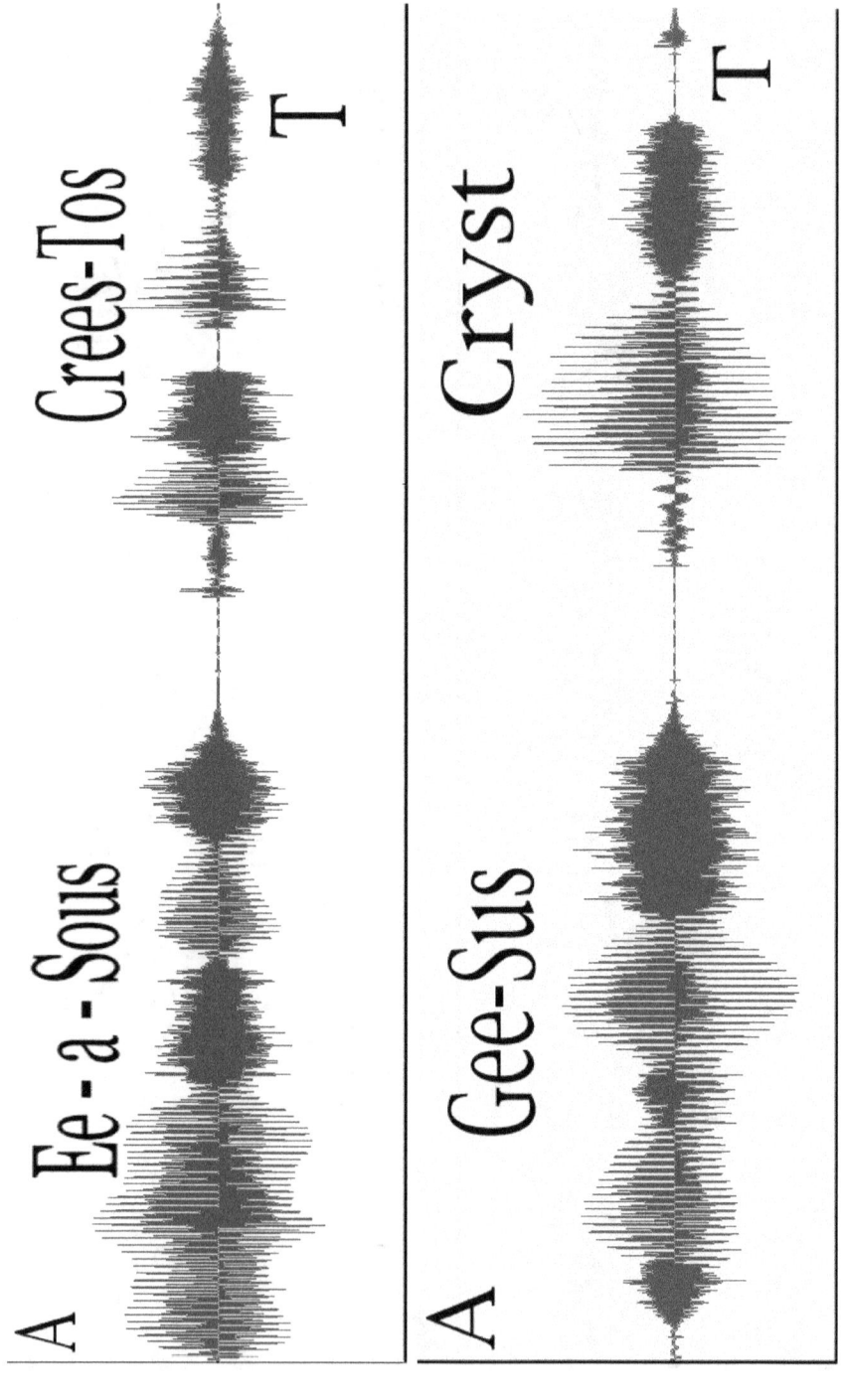

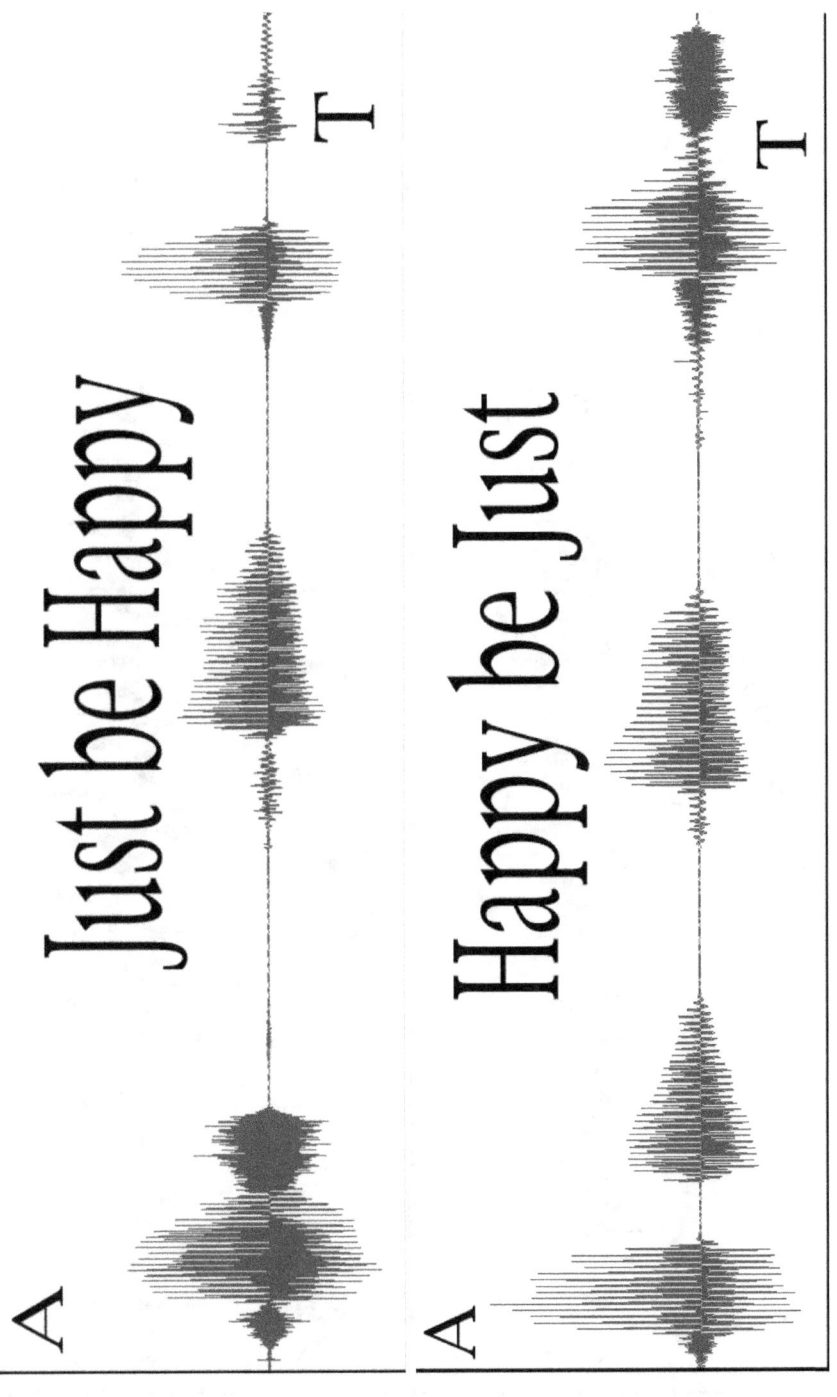

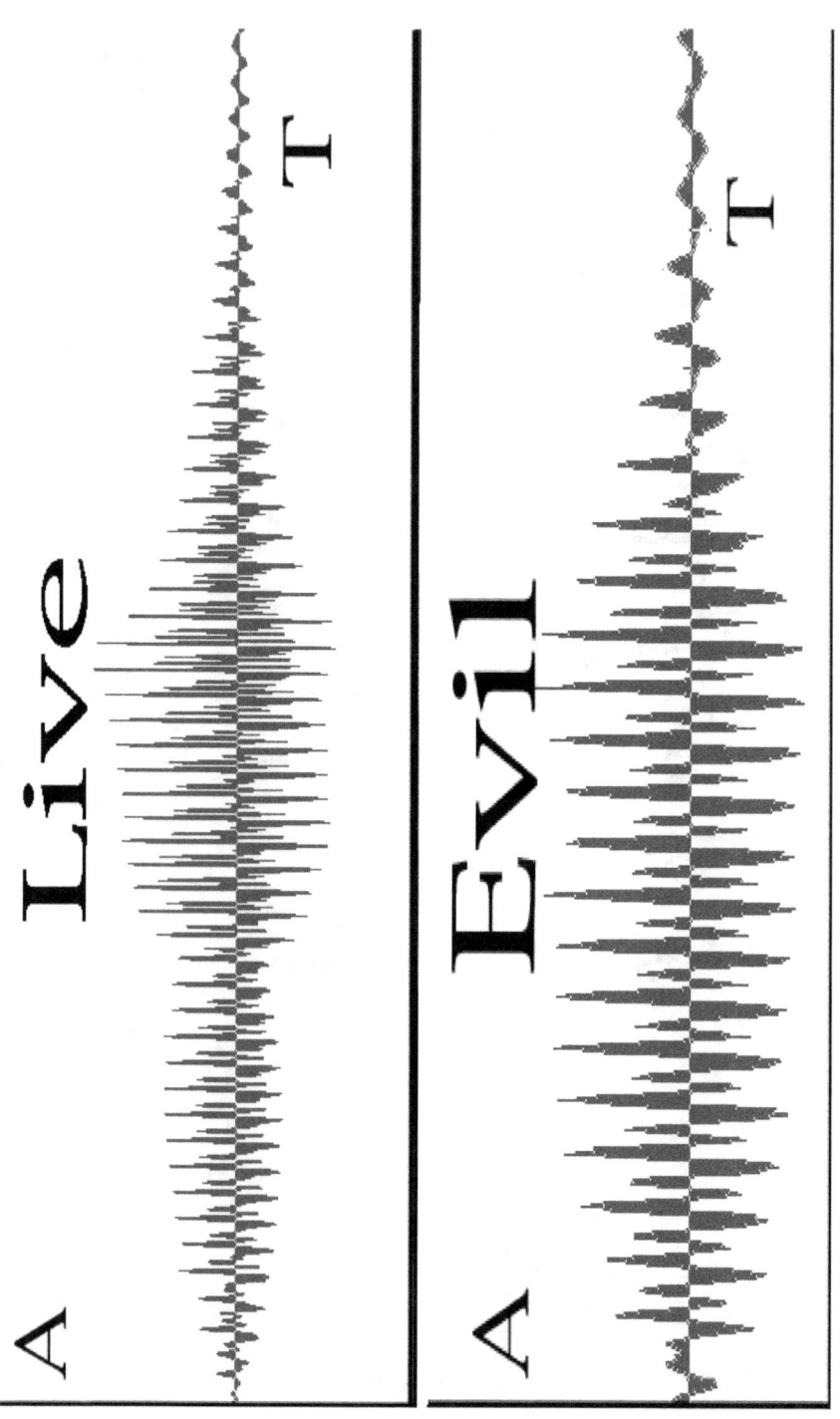

# YHSVH

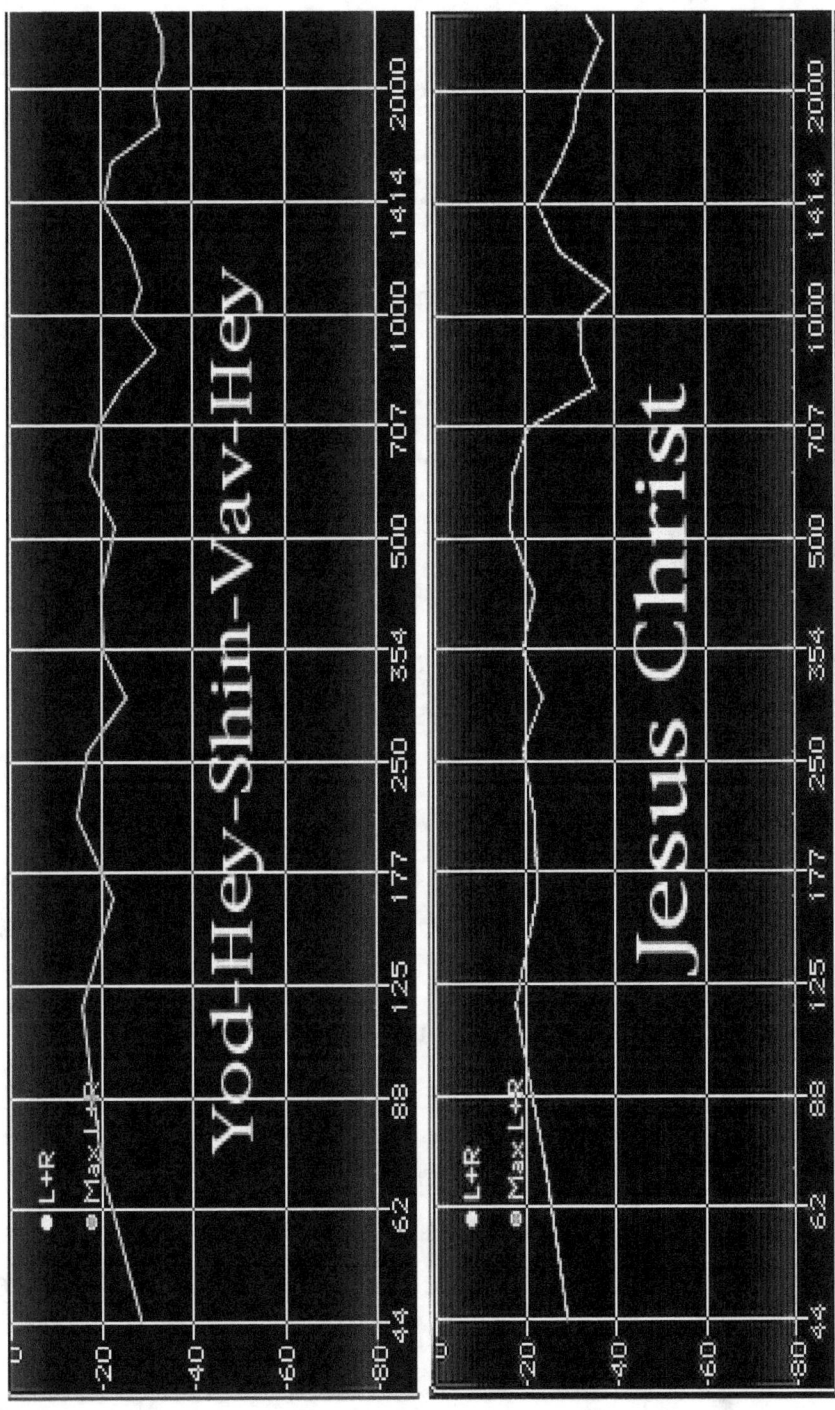

# YHSVH

Though very basic representations, the images on the preceding pages clearly illustrate that, <u>*despite your intent*</u>, different sound wave energy is produced when different words are spoken:

Page 58 – YHVH and Yahweh

Page 59 – Jehovah and YHSVH

Page 60 – Yashua and Iesous
The name Yashua is seen as an English spelling of YHSVH.

Page 61 – Jesus and YHSVH Mashiach
The term Mashiach means Messiah in Hebrew.

Page 62 – Iesous Christos and Jesus Christ

Page 63 & 64 – Here we see that when the same words, or letters, are spoken in a different sequence—a different sound wave will result.
(*see* page 52)

Page 65 – These are frequency (or tonal) graphs. Despite the intended meaning being the same—the sound frequencies are different. The vertical line is decibels; horizontal is kilohertz.

# YHSVH

In closing, some may be wondering—*Why such a fuss over the fact that different sounds produce "measurably" different waves?* Look, change the intent word—and you literally change the goal energy!7 Were creation a matrix of sound wave energy—this is tantamount to trying to put square pegs into round holes.

The scriptures teach—*Keep My name (My Sound) holy!* Wonder where changing it altogether fits into that equation? We also read: *"You may ask me for anything in my name, and I will do it."* If you want to experience His divine energy—do not be afraid to declare His divine name! Remember, *"Man does not live on bread alone, but on every word [divine sound] that comes from the mouth of YHVH."*8 I hope you will heed these words because *sound matters.*1

---

7 Note the different energy waves on pages 58 to 62.
8 John 14: 14 & Matthew 4: 4
Be well told—the source is always greater than the concoction. **Let no one get this twisted**, *This is no new teaching—this is only what you were told from the beginning.*

# Babel

## Babel

*"Therefore, I urge you, brothers . . . Do not conform any longer to the pattern of this world, but be transformed by the renewing of your mind. Then you will be able to test and approve what God's will is—his good, pleasing and perfect will."*

Romans 12: 1 – 2

## YHSVH

What have we come to understand? We have seen that sound waves can be extremely powerful and that all of these waves are not the same. Not only this, we have also learned that the pronunciation of the energy wave for the heavenly Father is YHVH (Yod-Hey-Vav-Hey) and that the energy wave for the Anointed One of the people who have come to be known as Christians is YHSVH (Yod-Hey-Shin-Vav-Hey).[1]

They didn't teach this to most of us in nursery school so I know that this may be difficult to fathom. That said, please do not shut your reasoning abilities down here *because in*

---

[1] Essenes, The Jewish Encyclopedia Vol. V, p. 231 & Christianity, Encyclopaedia Judaica Vol. V, p. 506 & The Riddle of the Dead Sea Scrolls (VHS) 1990 & Baptism, Encyclopedia of Religion Vol. II, pp. 60 – 61
The Essenes and early Christians referred to themselves as Followers of the Way. The term "Christians" was first applied to the Followers of the Way by pagans who described them as the "Callers on Christ." Eventually, the Christ Callers would become known as *Christians*.

# Babel

*desperate times people need to reason more—not less.*[1]

Actually, upon reflection, one cannot help but recall the story of the tower of Babel. In case you're not familiar with it, the scriptures explain that at one point people had one divine language. The fact that they had one language made it possible to achieve anything they set their minds to do. The people of Babylon desired to build a great and enduring tower. However, being displeased with their ambitions, YHVH confounded them by causing the people to speak different languages.

It is explained that the use of words that were not understood by all (*and that may well have not been divine*) made it impossible for the people to continue working on the tower—so it was abandoned and never completed. The tower became known as *Babel*, which is said to mean

"Confused." To this day, people who use many words but never seem to get to the heart of a matter are often seen as *babblers*.[2]

**Rendering of the Tower of Babel**

The reason that I have taken time to share this is because it is a cardinal mistake to disregard the importance of language. This is especially true regarding spirituality and the peoples of the West because of the nature of English. You see, of the tongues that linguists categorize as being

---

[2] Genesis 11: 1 - 9 & Veith, W., The Man Behind the Mask (DVD)
Babel literally means "Gate to God." Image is provided by Whimsy - http://karenswhimsy.com/public-domain-images.

# Babel

divinely inspired—Latin and English are not to be found. According to Hurtak, *Sacred Language* encapsulates the full experience and divine destiny. The Sacred Languages are Egyptian, Sanskrit, Hebrew, Tibetan and Chinese.

The importance of these languages is that they are vehicles for entry into a heightened state of consciousness with the Creator![3] The closer our consciousness comes to YHVH—the greater our access to His insight, power and authority. When the early followers of YHSVH were casting out evil spirits and healing the sick—they were unleashing YHVH's divine energy through the proper use of His divine sound waves! As the *Book of Luke* explains, *"The seventy-two returned with joy and said, 'YHSVH, even the demons submit to us in your name.'"*[4]

---

[3] Hurtak, J., The books of knowledge: the keys of Enoch p. 601 & Goldman, J., Holy Harmony (CD)
[4] Luke 10: 17

# 𝔜𝔋𝔖𝔙𝔋

| English | Akkadian | Aramaic | Hebrew | Ethiopic |
|---|---|---|---|---|
| Water | Mu | Mayya | Mayya | May |
| Sky | Samu | S-Mayya | Samu-yim | Samay |
| House | B-tu | Bayta | Bayl | Bet |
| Name | Sumu | S-ma | Sem | Sem |
| Head | Resu | Resa | Ro's | Re'es |
| Peace | Salamu | S-lama | Salom | Salam |

74

# Babel

As you can see, many of the sounds that are produced by the words of the sacred tongues and their corresponding words in English are vastly different. Another point here is that the Northeast African, Hebrew, Akkadian (Canaanite) and Aramaic languages are quite similar. The reason for this is because the first people to speak these tongues were originally of similar region, culture and race.

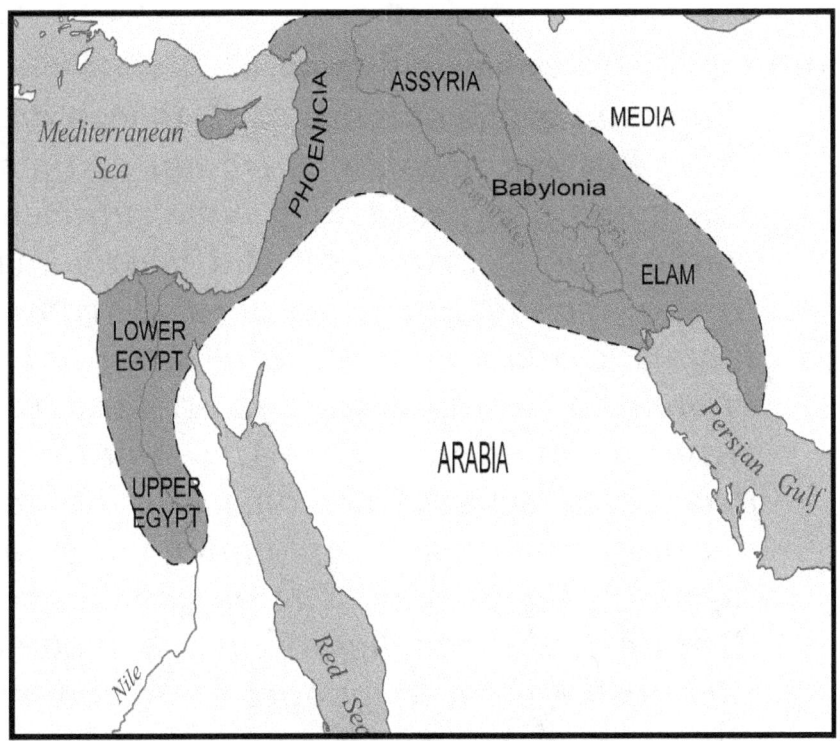

**Ancient Fertile Crescent**

# YHSUH

Here is a sampling of what linguists, historians and anthropologists have disclosed about the ancient people of the Fertile Crescent:[11]

Linguists explain that the ancient Egyptian and later Hebrew languages had hundreds of words derived from the same root words. According to Budge, *"The Egyptian alphabet has a great deal in common with the Hebrew and other Semitic dialects in respect of the guttural and other letters, peculiar to Oriental [Crescent] peoples."*

Additionally, Davies would remark:

> *"Egyptian is one of a group of African and Near Eastern languages (many of them still living tongues) which have sufficient similarities in grammar and vocabulary to suggest that they are derived from a linguistic common ancestor. This group is known to scholars as Afro-Asiatic (or Hamito-Semitic). The Afro-Asiatic is deemed at present to consist of six co-ordinate branches of which ancient Egyptian forms one. The other five are: Semitic (sub-branches of which include such well known languages as Akkadian, Hebrew and Arabic..."*

# Babel

**Noble Couple of Ancient Egypt c. 1400 BC**

Windsor says, *"The Phoenicians spoke a Hamitic-Semitic language so closely allied to the Hebrew that Phoenician and Hebrew, though different dialects, may practically be regarded as the same language . . ."*

Ackroyd and Evans have disclosed the following:
> *"Hebrew does not appear to have been the original language of the Hebrews themselves, but the language of the inhabitants of Canaan . . . it is more accurately described once in the Old Testament (Isa. 19:18) as 'the language of Canaan'. . ."*

What's more, academics have divulged the fact that the ancient Canaanite and Aramaic languages are so similar that many linguists flat out refuse to classify them as different languages. De Graft-Johnson states: *"The Phoenicians were Semitic people who, originally, appear to have resembled the Jews in race and language."*

In passing, the term *Oriental* was originally used by Western historians to designate the peoples of North Africa (often Egypt), the Southeastern Mediterranean, and Western Asia.

# Babel

**Phoenician man with goat and dog c. 900 BC**

## יהוה

The authors of *In the Shadow of the Pyramids* explain:

> "*Connections exist with ancient and modern Semitic languages of western Asia, as well as Cushitic, Berber and Chado-Hamitic languages of Ethiopia, Libya and the western Sudan. These, however, suggest a common origin rather than a superimposition of one language upon another.*"

Turning to Cox we find: "*The Kushite [Ethiopian] scripts were the origin of all Semitic scripts, and therefore the origin of the Akkadian (Babylonian-Assyrian) scripts, and of Hebrew, Aramaic, and the Mycenean [Minoan] scripts as well . . .*"

Chancellor Williams tells us that the Nubians (Ethiopians) created a written alphabet before the Egyptians! This 23-letter alphabet had 17 consonants, 4 vowels and 2 syllabic signs. In deference here to Gordon: "*The alphabet is the most useful single invention made by man throughout all history . . . The alphabet with such a limited repertoire of signs brought literacy within the grasp of whole nations and made universal education possible.*"

# Babel

**Ancient Nubian Man c. 1500 BC**

יהוה

Slouschz would observe the following about the famed North African metropolis of Carthage:

> *"We have found again the ancient language and writing of Canaan, the rich, idiomatic speech of a city which once counted seven hundred thousand inhabitants. And we Hebrew writers, we who write and feel in our Biblical tongue, have recognized at once that this re-called Phoenician language is nothing more nor less than Hebrew [and vice-versa] . . ."*

**Two Ancient Carthaginian Boxers c. 250 BC**

# Babel

**Elamite Soldier - Palace of Darius I c. 500 BC**

As is plain to see, the languages and peoples we have been discussing were not Caucasian. In truth, Black societies were thriving in this part of the world well after the life of YHSVH. In 484 A.D., the famous chronicler Moses of Khorene would characterize Western Asia thus: *"The name of Kush or Ethiopia applies to four great regions, Media, Persia, Susiana or Elam, and Aria [Northern Arabia] or to the whole territory between the Tigris and Indus [Southeastern Pakistan]."* (Map page 84)

**Map of Southwestern Asia c. 500 CE**

While many of these facts may be surprising to you, the fact that the sunny and warm regions of the Fertile Crescent were inhabited by Blacks in ancient times is hardly revelatory for most around the world. Over and above this, the noted anthropologist Calvin Kephart has gone to great lengths to show that the first Caucasian societies to establish themselves on the Northwesterly-most boundaries of Mesopotamia did not arrive there until centuries after the life of the Hebrew Patriarch Abraham. Even more telling, *the ancient Hebrews were to document the fact that they and the Elamites (see page 83) were of common ancestry!* Truth be told, many have trumpeted the racial identity of the Hebrews through the ages.

# Babel

This famous Greek portrait of the Hebrew Prophet Isaiah was painted in the 5th century. While it needn't be seen as an actual likeness of the great holy man—this markedly dark skinned depiction of a Hebrew prophet is probably as accurate as any ancient portrayal in existence. The much paler shading of the Greek (Indo-European) women is undeniable in the original portrait: a clear indication of the well-known racial distinction between the Hebrews and Caucasians. Image courtesy of Bibliotheque nationale de France.

## YHSVH

This type of iconography was not only found in the Mediterranean coastal regions. Channell explains that France has hundreds of sites where Black Madonnas are held in reverence. But nothing illustrates this fact more clearly than the 18th century testimony of Godfrey Higgins:

> *"In all the Romish countries of Europe, in France, Italy, Germany, &c., the God Christ, as well as his mother, are described in their old pictures and statues to be black. The infant God in the arms of his black mother, his eyes and drapery white, is himself perfectly black. There is scarcely an Church in Italy where some remains of the worship of the Black Virgin and the Black Christ Child are not to be met with. Very often the black figures have given way to white ones, and in these cases the black ones, as being held sacred, were put into retired places in the churches, but were not destroyed, but are yet to be found there. In many cases the images are painted all over and look like bronze . . . Their real blackness is not to be questioned for a moment."*

# Babel

Spain's Riccia painted this lovely Black Madonna in the 1600's. The Spanish treasure it to this day.

### 𝔜𝔥𝔖𝔳𝔥

I realize that old habits can be hard to break, but if we want to be more—we've got to be willing to do more of what's already been asked! WORD: <u>*Words matter*</u>. It's vitally important to heed Paul's warning in *II Corinthians* when he says:

> *"For if he who comes preaches another Messiah whom we have not preached, or if you receive a different spirit which you have not received, or a different gospel which you have not accepted—you may well put up with it!"*[5]

Try as one may, there can be no doubt as to what Paul's admonition is actually saying—*For if he who is to come later preaches another YHSVH (Yod-Hey-Shin-Vav-Hey) whom we have not preached . . . <u>you may well accept it</u>*.

This is the truth as clearly and concisely as I can bring it; thus, no more babel! I know that it will

---

[5] II Corinthians 11: 4

## Babel

be impossible for some of you to digest. Despite the fact that we are told to keep His name holy, it is as it was written: *"You have let go of the commands of YHVH (Yod-Hey-Vav-Hey) and are holding on to the traditions of men."*[6] Look, no one is asking you to become a linguist—but as we've clearly seen—*sound matters*. Therefore, let me suggest that a simple place to begin would be a short and familiar prayer because if you take one step—the Creator will take two . . .

> *YHVH who art in heaven, hallowed be thy name. Thy kingdom come, Thy will be done on earth, as it is in heaven. Give us this day our daily bread. And forgive us our debts, as we forgive our debtors. Lead us not into temptation, but deliver us from evil: For thine is the kingdom, and the power, and the glory, forever. In the name of YHSVH—Amen!*[7]

---

[6] Mark 7: 8
[7] Matthew Ch. 5 - 6

# Under the Sun

# Under the Sun

# JESUS

In 1929, an accomplished Russian scientist named Georges Lakhovsky wrote a book entitled, *The Secret of Life*. This work explained that every biological cell has the qualities of resistance, capacitance and inductance. Now what's so remarkable about this is that resistance, capacitance and inductance are most commonly discussed in regards to electrical components. Georges Lakhovsky writes:

> *"Every living cell is essentially dependent on its nucleus which is the center of oscillations and gives off radiations . . . These nuclei are . . . actual electric circuits endowed with self-inductance and capacity and consequently capable of oscillating. These circuits oscillate according to a range of wavelengths whose magnitude depends essentially on the values of spirals and capacities. The waves given*

*off are thus of electromagnetic origin . . . and are also of very high frequency and give off radiations of various frequency . . ."*[1]

Lakhovsky did not stop there however—he went on to try and discover a way to kill cancer cells. Lakhovsky's approach to the problem was to neutralize the cancer cells through the use of energy waves. In short, after discovering that cancer cells and healthy cells oscillate (move from point to point) at different frequencies— Lakhovsky decided that he would attempt to amplify the oscillation wave frequency of the cells that were healthy.

---

[1] Lakhovsky, G., Secret of Life: Cosmic Rays and Radiations of Living pp. 75 - 77 & Lynes, B., Cancer Solutions: Rife, Energy Medicine and Medical Politics p. 50
Lakhovsky would remark: " *Each cell is capable of being the centre of oscillations of very high frequency giving off invisible radiations belonging to a gamut close to that associated with light."*

What occurred was that boosting the amplification of the oscillations of the healthy cells literally overwhelmed the oscillatory energy of the cancer cells! Eventually, the weaker frequency producing cancer cells died off. Incredibly, Lakhovsky would reduce illness to nothing more than a conflict between the energy emitted from the oscillatory frequency of a healthy cell and that emitted by the frequency by a diseased cell.²

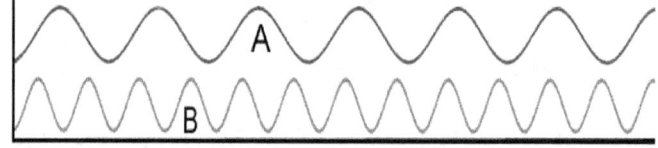

**This is an example of two different cells' oscillatory wave frequencies. If the healthy cell is oscillating at the top frequency (A), that's the frequency to be amplified over the unhealthy cell frequency (B) to destroy a cancer according to Lakhovsky.**

About 60 years ago, another Russian scientist would make a remarkable observation in regards to energy and life. During his studies of plants

---

² Lakhovsky, G., Secret of Life: Cosmic Rays and Radiations of Living Ch. VI, X & Lynes, B., Cancer Solutions: Rife, Energy Medicine and Medical Politics pp. 50 - 51

and their structures, A. M. Sinyukhin discovered that plants employ electrical energy during the healing process. Sinyukhin observed that when a plant looses a limb it generates an electrical current over the affected area that is maintained until the plant is healed. Further, Sinyukhin found that when he added a small electrical stimulus to a damaged plant—its healing process was accelerated over that of damaged plants that did not receive any outside electrical stimulus. <u>Over and above this, European researchers have recently discovered that human antibodies also use electrical charges to destroy the organisms which invade our bodies!</u>[3]

So, what has emerged is that electromagnetic energy plays a very important role in life at the molecular level. Question: *If life is dependent upon electromagnetic energy at its molecular*

---

[3] Becker, R., & Selden, G., <u>The Body Electric: Electromagnetism And The Foundation of Life</u> p. 61 & <u>The Beck Protocol</u> (DVD)

*level—what should we be employing to treat it when ill—a pill, a knife, or electromagnetic energy?* Pursuing the answer to this question brings us to Royal Raymond Rife.

In my estimation, despite the fact that most of you have never heard of him, Royal Raymond Rife might well have been one of the most significant figures in 20th century medicine. As a young man Rife decided that he would turn his love of machinery, invention and discovery towards making devices that would help combat disease. Becoming intrigued by the problems that virologists were encountering—Rife became convinced that the ability to see a virus was crucial to establishing how to destroy it.

During the decade of the 1920s, Rife would develop two amazing inventions. First, he would invent a very sophisticated electromagnetic wavelength frequency instrument that could be calibrated to precise frequencies. The other device Rife produced was a microscope that was capable of 17,000 magnification! *This was about 8 times better than the microscopes that most of the researchers of his day were using.* Royal Rife is recorded as being the first human to actually see a virus. After a conference attended by hundreds of doctors, the LA Times would describe Rife's invention as *"The World's most Powerful Microscope!"*

Next, Rife set out to destroy harmful viruses by employing the resonance principle previously discussed (pages 27 – 31). Rife theorized that if every virus depends upon electromagnetic energy and has a resonance point, then every virus can be destroyed. One simply has to present the proper electromagnetic wave (*radio wave*) to the virus to make it vibrate until it shatters—<u>and eyewitnesses explain that this is precisely what Rife's frequency machine did</u>!

### YHSUH

Royal Rife died in 1971 at the age of eighty-three. In the decades leading up to his death however, he would continue to make improvements to his frequency instrument and microscope. No less important, Rife would chart the M.O.R. (Mortal Oscillatory Rate) of many viruses. Once a virus's M.O.R. was established, Rife's frequency machine had no problem destroying it![4]

Many cancer patients are reported to have been successfully treated with Rife's energy wave device. Yet, he still had many detractors and a concerted effort has been made to keep his story from you. <u>Big Business and serious illness are a bad combination</u>. But in deference here to Rife:

> *"With the frequency instrument, no tissue is destroyed, no pain is felt, no noise is*

---

[4] Lynes, B., <u>Cancer Solutions: Rife, Energy Medicine and Medical Politics</u> pp. 117 - 118 & Bailey, D., & Wright, E., <u>Practical Fiber Optics</u> p. 23
The electromagnetic spectrum consists of radio waves, microwaves, terahertz radiation, infrared radiation, visible light, ultraviolet radiation, X-rays and gamma rays.

*audible, and no sensation is noticed. A tube lights up and 3 minutes later the treatment is complete. The virus or bacteria is destroyed and the body then recovers itself naturally from the toxic effect of the virus or bacteria . . . The first clinical work on cancer was completed under the supervision of Milbank Johnson, M.D. which was set up under a Special Medical Research Committee of the University of Southern California. 16 cases were treated at the clinic for many types of malignancy. After 3 months, 14 of these so called hopeless cases were signed off as clinically cured by the staff of five medical doctors and Dr. Alvin G. Foord, M.D. Pathologist for the group . . ."*[5]

---

[5] Lynes. B., The Cancer Cure That Worked! pp. 34 – 36, 43 – 47, 51, 60 – 61, 72 – 73 & Lynes, B., Cancer Solutions: Rife, Energy Medicine and Medical Politics pp. 47 – 48, 100 –103
Rife declared: "In reality, it is not the bacteria themselves that produce the disease, but the chemical constituents of these micro-organisms enacting upon the unbalanced cell metabolism of the human body that in actuality produce the

**Royal Raymond Rife's U.S. Patent for his High Intensity Microscope Lamp in 1929. No. 1,727,618**

---

disease. We also believe if the metabolism of the human body is perfectly balanced or poised, it is susceptible to no disease."

Coming to the 21st century, Dr. Len Horowitz reveals the importance of the relationship between electromagnetic energy and human DNA. In short, everything is made up of atoms. The atom represents the smallest known division of a chemical element.[1] Atoms are made up of three subatomic principles: protons, neutrons and electrons.

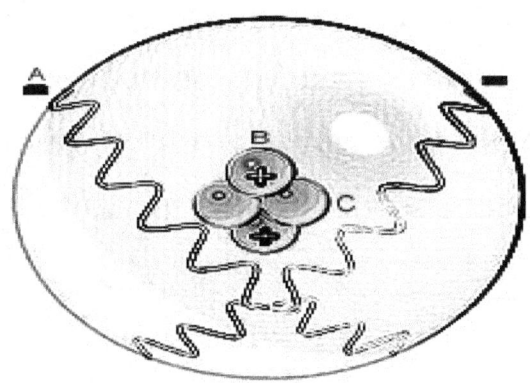

**Electrons (A) are particles that have a negative electrical charge which orbit around the nucleus (in the body) of the atom. Protons (B) are much bigger and have the opposite (or positive) charge. And Neutrons (C) are the same size as protons but they have no electrical charge.**

According to Horowitz, without the electromagnetically charged protons and

# YHSVH

electrons—atoms cannot exist. This is why scientists explain that electromagnetic energy and water are the true keys of life![6]

Turning to human anatomy, we are constantly told that our bodies are about 70% water and that water's most important function is to keep the body hydrated. In the most basic and wholly material sense this is correct; yet, this is only a part of the story.[7] In addition, we have been told that DNA simply houses our genes. In turn, those genes send out instructions to the rest of the body to determine such things as our height, skin color, eye color and so forth. Yet, once more, this is only a small part of the story . . .

---

[6] Johnson, R., Atomic Structure & Nardo, D., Atoms
It is the unique number combinations of these charged protons, neutrons and electrons that determines the properties of the atom: i.e., hydrogen, oxygen, Helium, etc., etc.
[7] Clayman, C., The Human Body: An Illustrated Guide to its Structure, Functions, and Disorders p. 175 & Heymsfield, S., Lohman, T., Wang, Z., Going, S., Human Body Composition pp. 35 - 46

What Dr. Len Horowitz's discloses in his *DNA: Pirates of the Sacred Spiral*, is that the principle function of DNA is to receive and transmit electromagnetic and sound wave energy! This is not that radical once stopping to reflect upon the earlier work of Lakhovsky, Sinyukhin and Rife.

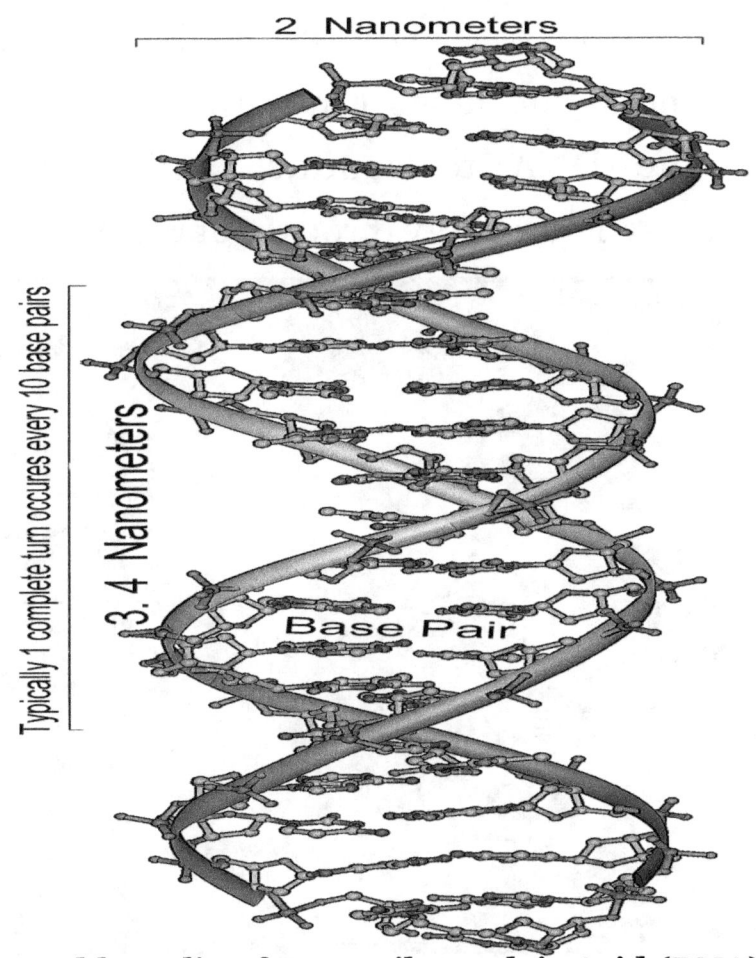

**Double Helix of Deoxyribonucleic Acid (DNA)**

Horowitz's goes on to reveal that less than 3% of your DNA is needed to determine your anatomical structure. Thus, the other <u>97% of your DNA is devoted to receiving and transmitting electromagnetic energy</u>. Moreover, while the process of hydration is obviously important, the other crucial functions of water are to encase and support the DNA strands—and help the DNA to conduct and transmit electromagnetic energy throughout the body!

**Concept of receiving divine energy rays c. 1400 BC**

Despite seemingly too incredible to believe, modern researchers have validated Horowitz's pronouncements. The sound frequency of 528 (*Mi* on the music scale) is said to help repair damaged DNA. It should be noted that Mi means *MI-Ra-Gestorum*, or "Miracle" in Latin. That notwithstanding, the crucial point here is that 528 (or any other frequency for that matter) could not repair DNA unless it possesses the ability to receive (or absorb) sound wave energy! According to the noted biochemist Dr. Lee Lorenzen: "*The 528 frequency is well known to scientists working on DNA repair.*"[8]

Continuing on, the helical formation of DNA (or *Sacred Spiral* as its called by Dr. Horowitz) has long been held by scientists to actually be an

---

[8] Calladine, C., & Drew, H., Understanding DNA: The Molecule & How IT Works p. 8, 13 & Horowitz, L., DNA: Pirates of the Sacred Spiral (DVD) & Horowitz, L., & Puleo, J., Healing Codes for the Biological Apocalypse p. 40 & Fativa: Ancient Healing Codes Revealed in Bible Are Published by Tetrahedron Press (Net) & Goldman, J., Holy Harmony (CD)

energy intensifying form. As a matter of fact, Nikola Tesla (one of the most brilliant physicist the world has known) understood this and employed the design in his heralded Tesla Coil!

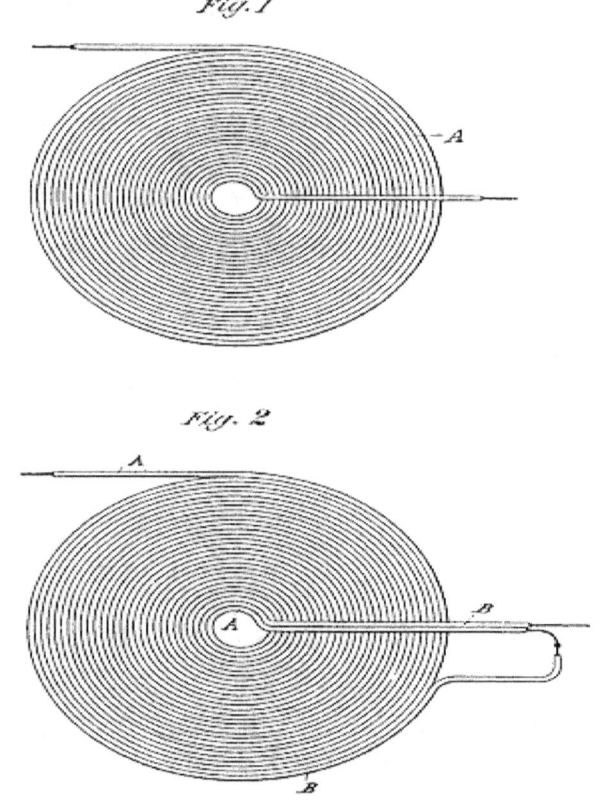

Diagram of the Tesla Coil U.S. Patent 512,340

The case is easily made that Nikola Tesla was also one of the world's finest electrical engineers. During his life, he would be issued more than 250 patents from 25 nations. Note this remark by Tesla in the U.S. Patent document for his electromagnetic coil:

> "I have found that in every coil there exists a certain relation between its self-induction and capacity that permits, a current of given frequency and potential to pass through it with no other opposition than that of ohmic resistance, or, in other words, as though it possessed no self-induction. This is due to the mutual relations existing between the special character of the current and the self-induction and capacity of the coil, the latter quantity being just capable of neutralizing the self-induction for that frequency . . ."[9]

---

[9] Twenty First Century Books: Coil For Electromagnets (Net)

Indeed, DNA's very helical design speaks to energy conduction! However moving from Tesla, I need to say something else about the electromagnetic nature of DNA. Calladine and Drew make the point that tiny electrical charges are actually crucial to the DNA strand's ability to maintain its helical (spiral) formation! According to our authors: *"The geometry at the core of the helix depends on subtle interactions between partial electrical charges . . ."*[10]

Finally, you will recall that Horowitz described our sacred spiraling DNA as transmitting and receiving physical energy. This helical design is also found in the hair of the Black Race. In actual point of fact, the hair of Black people is a series of (to use Horowitz's term) sacred spirals.

---

[10] Calladine, C., & Drew, H., Understanding DNA: The Molecule & How IT Works pp. 25 - 38
In passing, the double helix strands of DNA are found to twist in both directions—some right to left and others from left to right. The helical structure is not only found in DNA, it is also present in the body's RNA (Ribonucleic Acid).

# Under the Sun

In Africa to this day, people who grow their locking spirals long are often called *Messengers*.

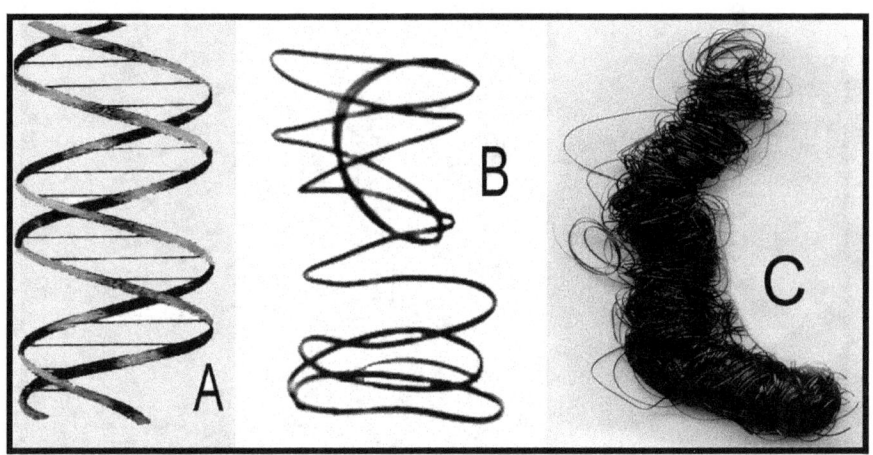

Image A is of the double helical pattern of DNA. Image B is of a single unadulterated (unstraightened and undamaged) Black hair (it is about 1/3 of an inch long). Image C shows a series of these naturally spiraling hairs gathering to form a new lock (it is about 1 1/2 inches long). When the Black hair shaft leaves the follicle it grows in the helical (or spiral) pattern. This tight ringlet form is in turn held by the activated melanin inside the hair. The ensuing images are of an Egyptian noble, two Black Babylonians, and an Egyptian goddess. The Babylonians did not typically wear their scalp hair in locks—but they often wore their beards in a locking fashion. However, the real point here is that the activated melanin contained within the helical hair shafts of a lock has long been held by biologists to be a receptor of ultraviolet light energy: *hence, another biological expression of (in the parlance of Tesla) the conductor and the coil!*[III]

# YHSVH

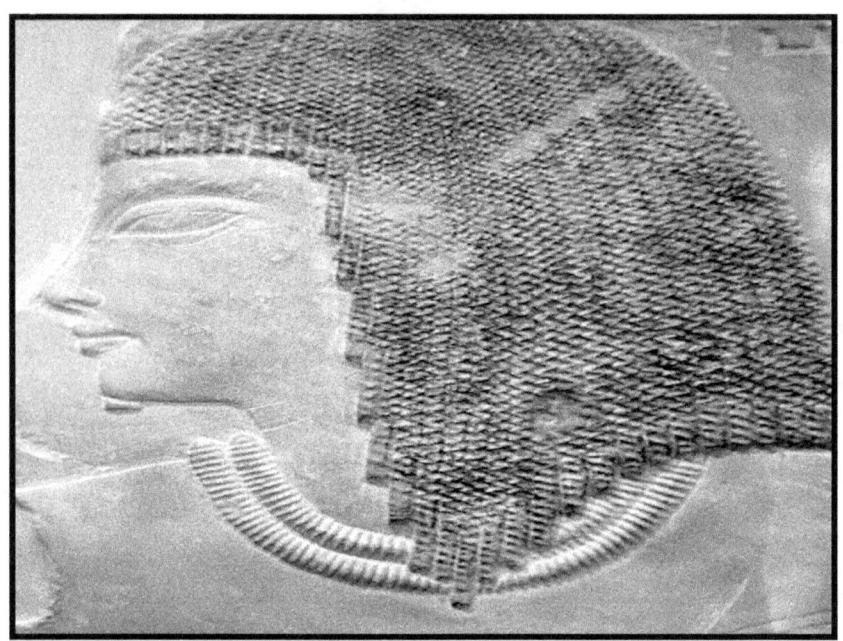

**Egyptian Noble C. 1400 BC**

**Ancient Black Babylonians of Mesopotamia**

# Under the Sun

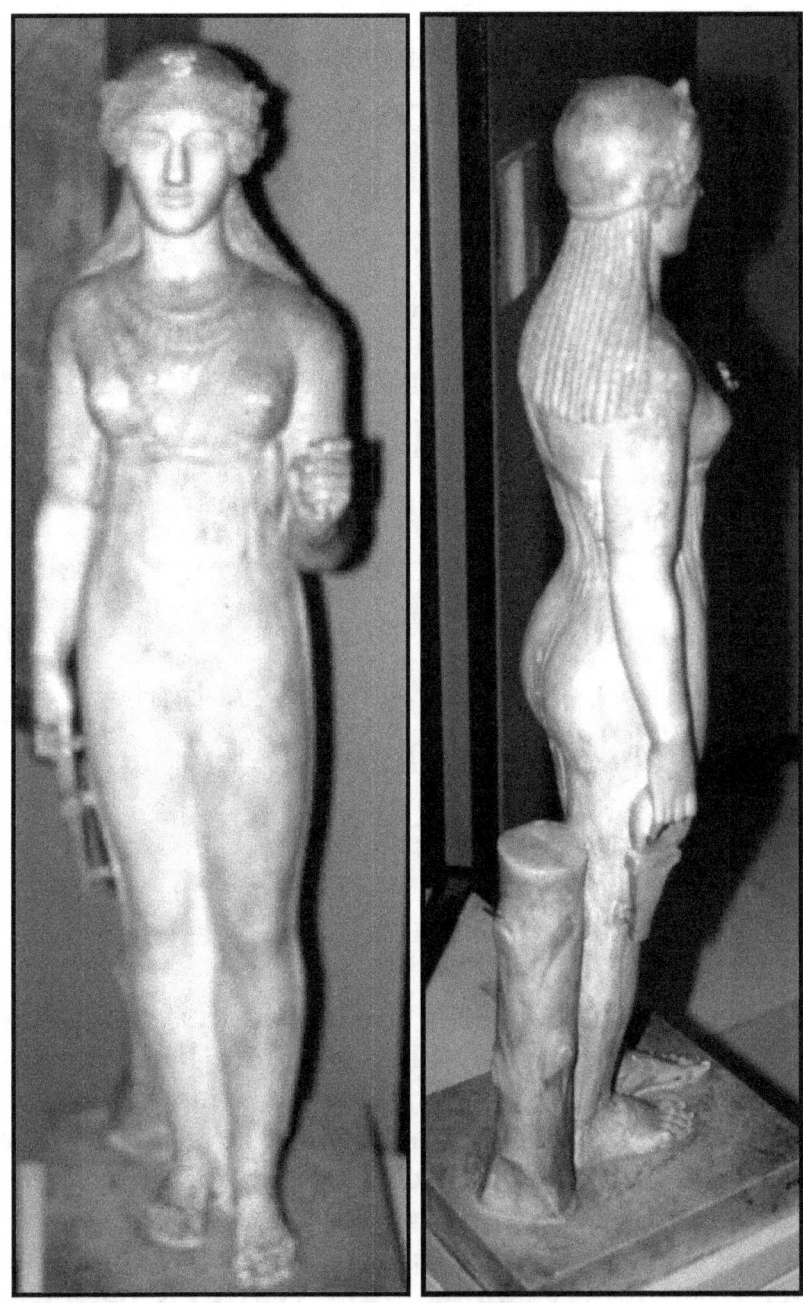

**This life size lock wearing Goddess represents the Mother of the Black peoples of the Nile c.300 BC**

So, bouncing from Lakhovsky, to Sinyukhin, to Rife, to Horowitz, to DNA's energy enhancing structure—what have we found?  Despite being told that our bodies consist of water and 6 passive elements—<u>we've not only found that every cell is electromagnetically charged—but our bodies are conducting symphonies of dynamic electrical functions daily</u>!  Hmmm', kind of makes sense once stopping to reflect upon the fact that when the heart stops—doctors don't prescribe aspirin or a cheeseburger—*they immediately send an electrical charge through the heart:* <u>CHARGING—**CLEAR**</u>*!!!*

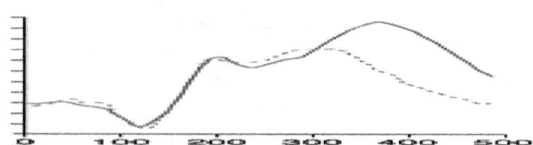

**Our brain activity is even measured in (you guessed it) electrical volts.  The vertical line denotes millivolts, while the horizontal line represents milliseconds. Look, it is a fact that energy waves are used on a daily basis to maintain and improve our health—from x-rays, to ultrasound, to EKGs. So why is there so much opposition to mild blood electrification? Wonder, if the reason is because it is safe, effective, and not monopolized by big business?** Image courtesy of NIAAA

## Under the Sun

Now, the reason that I took the time to share all of that is to explain this. Despite visiting scores of archeological sites and museums, reading many history books, and observing an even larger number of artifacts—there has been one prolific style of Egyptian sculpture whose meaning was not grasped by me until recently.

**Ancient Egyptian Man c. 2000 BC**

## 𝔜𝔥𝔖𝔲𝔥

Please note the two implements in the ancient Egyptian man's hands on the preceding page. They are most commonly referred to as the *Rods of Ancient Egypt* or the *Wands of Horus*. While having a fair understanding of the ancient culture, for so much as ancient texts about these items are not in abundance, these artifacts did not stir me to much consideration—before now.[11]

**These rods were made in various shapes; e.g., nonagon (9 sided), squares and circles.**

---

[11] Ions, V., <u>Egyptian Mythology</u> p. 60, 114 & Hobson, C., <u>The World of the Pharaohs</u> p. 134 & De Lubicz, R., <u>Sacred Science: The King of Pharaonic Theocracy</u> p. 237 & Tompkins, P., <u>The Magic of Obelisks</u> p. 449 & <u>Ancient Lives</u> (VHS) 1988

The reason that numerous texts discussing the rods have not been found will be discussed later. Horus had to overcome evil to receive spiritual jurisdiction over Egypt's kings. He was also the principal guide of the deceased king's soul in the Nether World. The Egyptians celebrated Horus as being, "The One Who Rights Wrongs!"

Although ancient documentation about the rods is scant, in that the ancients exhibited them so often, a great deal of considered speculation has been advanced by Egyptologists. For instance, some believe that holding the rods in your hands helps to heal the body by bringing its chemistry back into balance. It is said that the holder would use one style of rod for one illness and another for some other ailment.

It has also been advanced that the rods are a type of cosmic receiver—one rod representing the sun and one representing the moon. According to this theory, when holding the rods (which were thought calibrated to receive cosmic energy) the holder was brought into contact with the Creator's universal energy frequencies of health, intuition and wisdom.

There are other writers who feel that the rods were actually hollowed out and that charcoal and other elements were put inside of them to help bring the body's chemistry back into balance. And there are still other researchers who believe that the rods were not actually metal, but merely a cloth wrapping that was gripped in the hands.

**Pharaoh Menkaura and two goddesses c. 2500 BC**

Instead of wholly endorsing one of these theories, allow me to share an idea that others have not so far as I am aware. The better part of this chapter has been spent establishing the fact that the human body is electromagnetically charged and that it can be brought into balance and/or healed through the use of electrical energy. Allow me to use the rest of this chapter to explain why I believe that the ancient Egyptians were also aware of this fact!

Because so little has firmly been established, here is what I'm comfortable stating about the rods: (1) the rods were of such vital importance to the Egyptians that they would memorialize them; (2) the Egyptians associated the rods with divine power and communication; (3) the rods were metal that was held in the hand; and (4), those who used the rods were believed to experience good health and well being.

At the outset let me tell you that the Egyptians were not in habit of venerating useless objects. As a matter of fact, to advance a lie, or some groundless fantasy, was frowned upon. Indeed, the ancient Egyptians taught, *"Lying is an abomination!"* Thus, you can rest assured that the memorialization of these rods by the Egyptians was no less considered than the science conveyed through the pyramids: put another way—*these instruments were for real!*

**Noble Egyptian Couple c. 2000 BC**

## Under the Sun

In my opinion, the energy in your body that 21st century scientists refer to as *Electromagnetic*, and that many of you consider your *Spirit*, was known to the Egyptians as the *Ka* and *Ba*. Suffice it here to say that the Egyptians considered the Ka and Ba to be two facets of the soul, which emanated from the Creator.[12]

---

[12] Fischer, H., <u>An Elusive Shape within the Fisted Hands of Egyptian Statues</u> & Uvarov, V., <u>Wands of Horus</u> & Mc Coy, R., <u>Under the Tamarisk Tree</u> & Gunn, B., <u>The Instruction of Ptahhotep and the Instruction of Kegemni: The Oldest Books in the World</u> & Soul, <u>Encyclopedia of Religion</u> Vol. XIII, pp. 432 – 434 & Bunson, M., <u>A Dictionary of Ancient Egypt</u> pp. 1, 10 – 11, 41, 130 & Butzer, C. (Ed.), <u>Ancient Egypt: Discovering Its Splendors</u> pp. 156 - 157 & Mertz, B., <u>Red Land, Black Land</u> pp. 288 - 296 & Diop, C.A., <u>The African Origin of Civilization: Myth Or Reality</u> pp. 190 - 191 & Budge, E.A., <u>Dwellers on the Nile</u> pp. 257-275 & Budge, E.A., <u>Egyptian Religion</u> p. 214 & Moscati, S., <u>The Face of the Ancient Orient</u> p. 127 & Budge, E.A., <u>A Short History of the Egyptian People</u> p. 225 & Sykes, E., <u>Everyman's Dictionary of Non-classical Mythology</u> p. 1, 26, 102, 113, 117 & Budge, E.A., <u>Osiris: The Egyptian Religion of Resurrection</u> Vol. II, pp. 117 - 118 & Muller, M., <u>Mythology of All Races</u> Vol. XII, p. 171 & Montet, P., <u>Eternal Egypt</u> p. 167 & Ions, V., <u>Egyptian Mythology</u> p. 104 & Kaster, J., <u>Wings of the Falcon: Life and Thought of Ancient Egypt</u> p. 206 & Budge, E.A., <u>The Book of the Dead</u> p. 259 & Perl, L., <u>Mummies, Tombs and Treasure</u> pp. 13, 14 - 15, 48 & Hall, M., <u>The Secret Teachings of All Ages</u> p. CLXXXIII & Irram, S., <u>Death and Burial in Ancient Egypt</u> p. 24, 31

Briefly, it should be noted here that scientists have actually given credence to the existence of electromagnetic life force (*Spirit* or *Ka* and *Ba*) energy. In 1929 Lakhovsky wrote:

> *"Let us observe the fact . . . that dead plants and animals do not give any evidence of detectable radio-activity, for it appears that natural radiation is essential . . . for the maintenance of life . . . These observations . . . have enabled Nodon to come to the following conclusion: 'It appears . . . that vital cells of the human body emit electrons generated by an actual radio-activity whose intensity would seem to be much more considerable than that observed in insects and plants. The fact that there should be a certain emission of energy in living beings . . . can hardly be doubted."*[13]

---

[13] Lakhovsky, G., <u>Secret of Life: Cosmic Rays and Radiations of Living</u> pp. 80 - 81 & Horowitz, L., <u>DNA: Pirates of the Sacred Spiral</u> (DVD)

But to return to the historic significance of the wands—as stated, blood electrification is a process that immobilizes viruses, parasites and bacteria. It is achieved by non-invasively sending a tiny electrical charge through the blood stream. Clinicians have widely heralded the health benefits of blood electrification.[14] Now while my rationale will in part be based upon the anecdotal *(for the previously stated reasons)* bear with me as I prove that these rods were tools of ancient blood electrification.

It is my view that the rods were hand-held conductors of electricity. As the direct current energy began to flow, the invisible Ka and Ba energy would immediately be felt entering the body through the hands. Obviously, it would only be natural for the Egyptians *(w h o*

---

[14] Becker, R., & Selden, G., The Body Electric: Electromagnetism And The Foundation of Life & The Granada Forum: Suppressed Medical Discovery (DVD)
Blood electrification immobilizes viruses be they man-made or not. Also see pages 43 and 44 in this text.

𝔜𝔋𝔖𝔙𝔋

*according to Herodotus were the most spiritual people in the world*) to declare this invisible energy an aspect of the creator (*invisible things being associated with the eternal*). Further, as those who were ill would become better (*from the destruction of invading organisms*) the rods would be seen as vital instruments: holy devices for the transmission of invisible life force (*Ka and Ba*) energy into its user![15]

Also, though again admittedly anecdotal, corroboration for blood electrification comes to

---

[15] Burn, A., & Selincourt, A., Herodotus: The Histories pp. 143, 151 - 152 & Murphy, E., Diodorus on Egypt & (Audio), Ancient African Medical Practices C. Finch 1992 & David, R., The Egyptian Kingdoms p. 114 & Casson, L., Daily Life In Ancient Egypt & Sauneron, S., Priests of Ancient Egypt & II Corinthians 4: 18 & The Granada Forum: Suppressed Medical Discovery (DVD)
Herodotus lived c. 450 B.C. He explained that the spirituality and philosophical understanding of Greeks was elementary in comparison to that of the Egyptians. Incidentally, it was widely held throughout the Old World that invisible things were of the heavenly realm. For instance, the Apostle Paul wrote: *"So we fix our eyes not on what is seen, but on what is unseen. For what is seen is temporary, but what is unseen is eternal."*

us through the way the wands are *consistently* portrayed. Firmly gripped wands in the hands (often while in a sitting, or still, position) to perform blood electrification makes perfect sense: (1), because it is an easy way to make, and keep, contact with the rod; (2), the arteries in the hands are relatively near the surface of the skin, which means that the electric current would not have to penetrate much tissue or muscle to enter the blood stream; and (3), with the sensation of the DC (Ka/Ba) energy, one would want to be in a sitting or rigid position.

Query: If an item facilitates invisible energy's entry into the body and the result is to feel better and live longer than others who do not use it—Could such a device be described as miraculous or *"magical"* like say—a <u>Wand</u> of <u>Horus?</u> The answer is only too obvious. Moreover, it goes without saying that any ancient device capable of achieving this would be widely commemorated—*just like we have observed throughout the dynastic history of ancient Egypt with the Wands of Horus . . .*

**Egyptian noble gripping one wand c. 2000 BC**

So, what we have is a revered item that (through physical touch) brings its user into direct contact with divine (invisible) energy and also produces a healthier state of being![II]  <u>Save the ancient spiritual proclamation and this describes modern day blood electrification precisely</u>!

# Under the Sun

**Inspector of Scribes
Raheka & Wife Merseankh c. 2300 BC**

A point that can't be stressed enough is that the Egyptians were empirical practitioners. They were scientists, physicians, surgeons, embalmers and mummifiers of the first order! As such, observing the health differences between those who used the wands, and those who did not, would have been rudimental. For example, Beck explains that blood cleansed by electrification and placed on a slide can take months to loose viability and crenate. *Conversely, non-electrified blood placed on a slide will typically crenate in less than a week!* In truth, the root word for the modern term <u>Chemistry</u> was *Khemet*, the ancient African name for Egypt.

### 𝔜𝔋𝔖𝔙𝔋

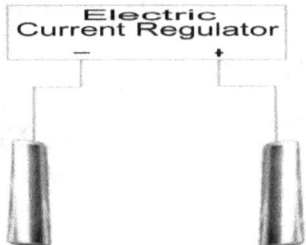

This is a generic illustration of a modern day energy device. The basic operating principles behind it are very much in line with the artifacts of the Egyptians. Here we have a current regulator. Next, we have positive and negative terminals that are connected to wires that are connected to two metal energy conductors (small rods in this case). It should be noted that, of course, the Egyptians used DC not AC electricity. <u>ONE POINT HERE</u>: *This image is not an endorsement of any kind of product—it is simply shared to demonstrate the similarity in configuration between such energy devices today and those of the past. Modern researchers extol the benefits of blood electrification. These appliances can be made or purchased for a minimal expense. However, <u>you must educate yourself about the products and methodologies to avoid harm and/or wasting money</u>. You might start with "<u>Suppressed Medical Discovery</u>" by The Granada Forum, which features <u>Dr. Robert C. Beck</u>.*

  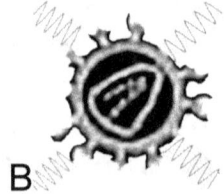

A                                    B

Image A represents a healthy virus. B denotes the same virus being exposed to <u>50 - 100 microamperes</u> of electric current (see pages 43 – 44).

## Under the Sun

**Pharaoh Rameses II c. 1200 BC**

**Rameses II lived to be very near <u>one hundred</u> years old. While every pharaoh didn't live to that age, it was not uncommon for Egyptian nobles to live well into the sixties, seventies, eighties and beyond!** <u>Now what's so remarkable about this is that they had achieved this in times when most peoples on earth were not living beyond their late thirties (see pp. 184 - 188). Let me submit that atop the considerable medical prowess of the ancient Egyptian physician was the practice of blood electrification: a proven health procedure that's being heralded today</u>**!**

## YHSUH

At this time, I want to touch upon a few issues that might well be problematic for you insomuch as we've discovered that—*You Can Fathom The Truth!* First, it is true that some writers think that these rods were not metal because some artifacts seem to indicate the presence of cloth in the hands holding the rods. If true, the wands would not be capable of conducting electricity.

I believe that the simplest way to verify that the rods were metal has already been touched upon. Return to page 114 for a moment and look at the shapes of the rods. Many of the existing artifacts reveal hard geometric sides. This could not happen if the only thing in the hand was a cloth wrapping. These well-defined flat surfaces are a clear indication that the rods were most likely metal. Also, just in case you aren't aware of it, the Egyptians were master metallurgists.[16]

---

[16] Element, Chemical, The World Book Encyclopedia Vol. V, p. 2280 & Iron, Metallurgy of, Universal Standard Encyclopedia Vol. XIII, p. 4436 & Van Sertima, I., Blacks in

# Under the Sun

**Gold Coffin of Tutankhamen c. 1350 BC**

---

Science: ancient and modern  p. 9 & Trigger, B., Kemp, B., O' Connor, D., & Lloyd, A., Ancient Egypt: A Social History & Malek, J., & Forman, W., In the Shadow of the Pyramids p. 31 Daniel, G., Nubia Under the Pharaohs pp. 66 - 67 & The World Book Encyclopedia Vol. V, p. 2228d, Vol. X, p. 435 & Edwards, I., The Pyramids of Egypt p. 250 & Budge, E.A., The Dwellers on the Nile pp. 37, 60 - 61

We find gold, silver, copper, tin and iron being mined and smelted at a very early date in Africa. Scholars write: *"A number of elements were in use before the dawn of history. Among these were copper, tin, lead, gold, silver, iron, and sulfur. In general, such well known elements are the ones that are found in a pure form in nature or which can be separated from their ores at relatively low temperatures."*

## JESUS

However, let me say this about the cloth assertion. I believe that those who have observed this may not be entirely wrong. What they are seeing may be cloth wrapped around a metal object that's being firmly gripped (e.g., possibly implied on page 113). The reason why is because, then as now, moist cloth could have been used to reduce skin irritation caused by the electrical energy entering the body.[17]

Regarding the actual methodology for using the wands, let me suggest that there could have been

---

[17] Petrie F., Wisdom of the Egyptians p. 19 & Budge, E.A., Osiris: The Egyptian Religion of Resurrection Vol. I, p. 388 & Muller, M., Mythology of All Races Vol. XII, pp. 54 - 55, 388 - 389 & Butzer, C. (Ed.), Ancient Egypt: Discovering Its Splendors p. 155 & Waddell, W., Manetho, Ptolemy pp. xxiii - xiv, 193 & Randolph, P., Hermes Trismegistus: his Divine Pymander p. 37 & Hall, M., The Secret Teachings of All Ages pp. LII, XCIX - C & Waddell, W., Manetho p. 191 & Fernie, W., The Occult and Curative Power of Precious Stones & DeGivry, G., Witchcraft, Magic & Alchemy
In passing, because Horus was seen as a solar (as opposed to lunar) deity, gold which was also considered to be a solar metal may have been employed in the making of these implements. Gold also just happens to conduct electricity better than any other metal.

two ways.  First, a tightly held rod could simply have been placed in contact with the charged metal of a DC battery and the energy would immediately flow.  Or, if the rods were designed with a hollowed out compartment (as some have suggested) they could have been filled with the proper elements needed to make them function as DC batteries while being firmly gripped.[18]

Oh, before I forget, I must address the most pivotal question many of you have; namely, how can I take the position that the ancients knew about electricity? *Needless to say, I'm sure that a lot of you are thinking that I would have to be crazy to think that!*  Especially because everybody knows that electricity was not known to humans before the lives of such heralded Europeans as Robert Boyle, Stephen Gray,

---

[18] Permit me to make the point here that were you to accept my view—each of the earlier theories advanced by Egyptologists (pp. 115 - 116) could now be somewhat generally applied to the rods.

### JESUS

Benjamin Franklin and Alessandro Volta! *Ergo, <u>How Could The Ancient Egyptians Possibly Know Anything About Electricity</u>?* Well, what we think we know, and what happens to truly be so, are often two entirely different matters . . .

**Count Alessandro Giuseppe Volta c. 1820 CE**

One of the simplest ways to prove that the ancient Egyptians had an understanding of electricity is by delving into another glaring question—*How did the Egyptians decorate, never mind construct, the exquisite interiors of their tombs and enclosed temples?* The answer is—*By using electricity to light them!*

# Under the Sun

**Entry way to a tomb in the Valley of the Kings**

**Interior wall in the Temple of Nefretari c. 1200 BC**

## 𝔜𝔥𝔖𝔳𝔥

Modern archeologists and scientific researchers have established the fact that the ancients were the inventors of primary cell batteries, which produced direct current electricity![19]  Briefly, archeologists could not explain the discovery of ancient jewelry with one metal melded to, or coating, another (often, a silver item with a gold outer covering).  To this day, the process for achieving this involves fusing the metals together by passing electricity through them; the process is known as *Electroplating*.

Then, in 1939 Wilhelm Konig (Director of the Nation Iraq Museum) observed an ancient artifact that he believed to be the key to the ancient electroplating question. What he found was an earthenware jar (A) with an asphalt (or tar-like) lid.  The underside of the lid had a hollow copper tube (B) connected to it.  The lid

---

[19] Tompkins, P., The Magic of Obelisks p. 456 & Berlitz, C., Atlantis: The Eighth Continent  pp. 128 - 129  &  Electric Battery, The World Book Encyclopedia  Vol. V, p. 2248

also had a thin lead rod (C) running directly through its center that went deep into the jar and protruded out of the top of the jar.

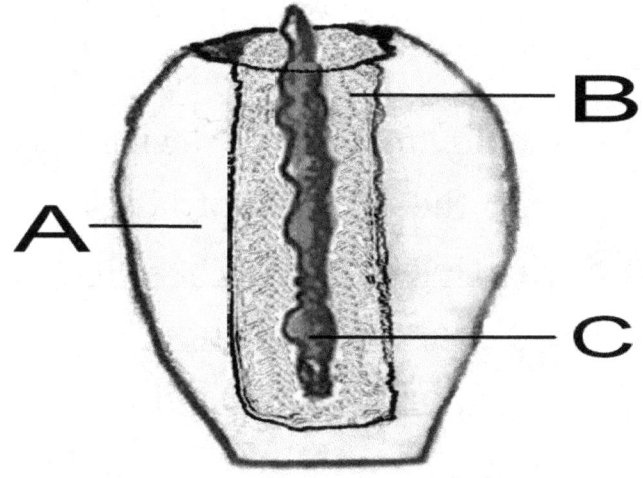

**Diagram of the Baghdad Battery**

Believe it or not, <u>all that was needed was an acidic liquid for this ancient artifact to produce electricity</u>! Scientists explain that there is evidence from chemical analysis that the ancients actually used lemon juice, grape juice and/or vinegar in the device. The artifact has come to be known as the *Baghdad Battery*. After testing the object, Dr. Arne Eggebrecht and Wilhelm Konig declared that by just adding an

𝔜𝔥𝔖𝔳𝔥

electrolyte (grape juice in their case) DC electricity was produced! In other tests of this type of battery, researchers produced more than 4 volts of electricity from softball-sized batteries.

In point of fact, many of these devices have been found amongst the artifacts of kings. <u>Add the finds of ancient electroplated jewelry in Egypt and all doubt that the ancients did not know about electricity, or have the ability to electrify blood, is rendered moot</u>! In passing, despite the fact that the ancients discovered the technology millenniums ago, this style of battery is known today as the *Voltaic Cell Battery* in honor of the Italian physicist Conte Alassandro Giuseppe Volta who lived during the 19th century CE.[20]

---

[20] <u>Arthur C. Clarke's Mysterious World: Ancient Wisdom</u> Vol V. (VHS) 1980 & Berlitz, C., <u>Mysteries From Forgotten Worlds</u> p. 24 & <u>Ancient Electricity</u> (net) & Berlitz, C., <u>Atlantis: The Eighth Continent</u> pp. 128 - 129 & <u>The World's Strangest UFO Stories</u> (VHS) & Plim Report: <u>Did Ancient Cultures know about Electricity?</u> (net) & <u>Electronics 101: Fundamentals of Electricity</u> (net) & Electric Battery, <u>The World Book Encyclopedia</u> Vol. V, p. 2248, Vol. XX, p. 446

## Under the Sun

With proof of ancient electricity production, the next question is—*Did the Egyptians possess the metallurgy skills to conduct it and create light?* As touched upon previously—*they most certainly did!* According to De Lubicz: "*Perfection was attained in copper workings and encountered in a vast variety of objects.*"[21] Copper is one of the best-known conductors of electricity; it is second only to gold.

The only other matter to address here then is—*Did the Egyptians understand how to create glass?* The fact of the matter is that the peoples of the Nile were making glass as early as the 4th millennium BC. In deference here to David:

"*Egyptian craftsmen became increasingly skilled during the Archaic Period (circa*

---

[21] De Lubicz, R., Sacred Science: The King of Pharaonic Theocracy p. 114 & Ancient Lives (VHS) 1988
In an attempt to illustrate the relative metallurgy capabilities of the ancient Egyptians, let me share this. When modern museum curators attempted to re-polish the golden sarcophagus of the famed Tutankhamen—the best methods known to them only scratched and dulled the gold work of the Egyptians.

*2800 B.C.). They produced magnificent stone vessels and experimented in working with faience and blue glass . . ."*[22]

**This is an ancient Egyptian blue glass artifact
It is about 36 inches long by about 10 inches wide**

Egyptologists explain that the ancient Egyptians were also amongst the first people to blow glass. The ancient Phoenicians were also adept at glass blowing. However, the oldest surviving specimen of ancient Egyptian glass dates to c. 1450 BC (about 3500 years ago).

---

[22] David, R., The Egyptian Kingdoms p. 14 & Casson, L., Daily Life In Ancient Egypt pp. 20 - 21 & Cox, G., African Empires and Civilizations p. 26 & Mertz, B., Red Land, Black Land p. 104 & Caxton's History of the World Vol. I & Sewell, P., Egypt Under the Pharaohs & Egypt, The World Book Encyclopedia Vol. V, p. 2228d & Malek, J., & Forman, W., In the Shadow of the Pyramids p. 55
A vacuum can be created by simply burning a candle inside of a glass turned up-side down until the oxygen is consumed.

This rendering is of a scene in the Temple of Hathor at Dendera. <u>This depiction (along with several others) has led many scholars to unequivocally declare that the Egyptians lit their tombs with large electric light bulbs</u>! Engineers even believe that they used heavy cabling and wiring to conduct electricity. Incidentally, the celestial metal of Hathor just happened to be copper. Yet, after a careful examination of the depictions in the temple at Dendera—Dr. John Harris of Oxford University would remark: *"The cables are virtually an exact copy of engineering illustrations as currently used."*[23]

---

[23] Clark, R., <u>The Sacred Tradition in Ancient Egypt</u> p. 158 & Berlitz, C., <u>The Bermuda Triangle</u> pp. 162 - 163

**Wall scene in the Temple of Hathor at Dendera**

Tompkins explains that the Egyptian word for electricity was <u>*Aor*</u>, which means "<u>Magic Light!</u>" *Hmmm'*, <u>*if magic and electricity were synonymous—then you might say hand-held "wands" and ancient blood electrification kind of go hand-in-hand!*</u>

---

[24] Tompkins, P., <u>The Magic of Obelisks</u> p. 456
This photo is of one of the ancient temple's chambers. It appears courtesy of Mr. Lasse Jensen and Wikimedia Commons - CC-BY-2.5.

As if this wouldn't be enough, in his *The Fantastic Inventions of Nikola Tesla,* the great inventor advances his belief that the ancients transmitted energy by the design and placement of their ancient pyramids and obelisks.

**This is a scale replica of the Giza Pyramid Complex of the ancient Egyptians. It was constructed c. 2500 BC.**

A point that should be made is that the pyramids we see today are different from the structures that were completed by the ancient Egyptians. Chroniclers tell us that the huge limestone blocks visible to us were completely encased by highly polished and perfectly placed stones.

# JESUS

Scholars explain: *"Outer casing stones were installed from the top down . . . These glistening, highly polished stones were covered with inscriptions—later lost when the blocks were carted off to Cairo."*[24]

Whether Tesla is correct or not about this energy transmission—as we know that the Egyptians produced electricity—one wonders why their writings do not discuss it more often? Let me submit that there are two reasons: First, because of the dastardly actions of the later Greeks and Romans; i.e., the sacking of Egypt's royal libraries and the burning of thousands of books! For instance, Diocletian burned all of the Egyptian books on the gold and silver crafts that he could. And secondly, the Egyptians were understandably not in the habit of sharing their

---

[24] Tesla, N., & Childress, D., The Fantastic Inventions of Nikola Tesla pp. 279 - 284 & Mystic Places p. 49 & Nova: Pyramids: Interview with Dr. Zahi Hawass, Director of the Pyramids (Net)
Dr. Hawass believes that the pyramids had gold cornerstones.

divine secrets with foreigners. The ancient Greek historian Strabo reports:

> *"The Egyptian priests are supreme in the science of the sky. Mysterious and reluctant to communicate, they eventually let themselves be persuaded, after much soliciting, to impart some of their precepts; although they conceal the greater part..."*[25]

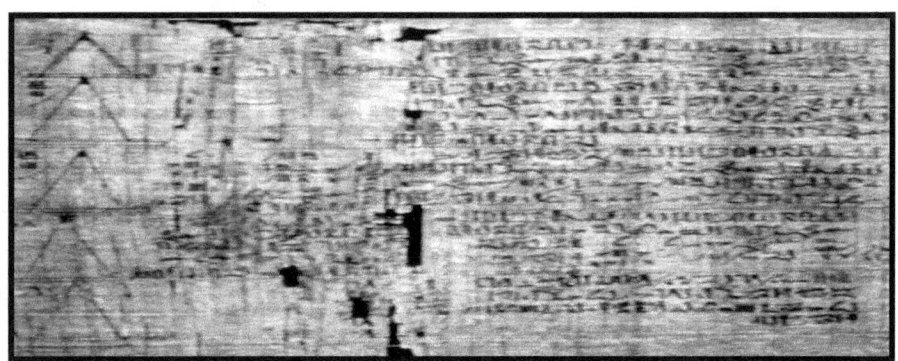

**This is a portion of the celebrated Rhind Papyrus c. 1700 BC. The papyrus demonstrates the higher mathematical capabilities of the ancient Egyptians.**

---

[25] Butzer, C. (Ed.), <u>Ancient Egypt: Discovering Its Splendors</u> p. 80 & Berlitz, C., <u>Mysteries From Forgotten Worlds</u> pp. 35 - 36, 80 - 81 & <u>Mystic Places</u> p. 49 & Berlitz, C., <u>Atlantis: The Eighth Continent</u> p. 129 & Dampier-Whetham, W., <u>A History Of Science: And its relations with Philosophy & Religion</u> pp. 51 - 52 & Bauval, R., & Gilbert, A., <u>The Orion Mystery</u> pp. 182 - 183

*In closing, not only have we found that the ancients were aware of unseen energy, they were utilizing it in ways that we can only imagine! This notwithstanding, from electricity to blood electrification, to the light bulb—one understands why philosophers down through the ages have made the pronouncement—*

**There's nothing new under the sun . . .**

# Epilogue

## YHSVH

I realize that we've touched upon a lot of physical and spiritual material—but it was necessary insomuch as real life is not a fractional proposition. From sound waves to blood electrification, I've tried to introduce you to the gains to be made by the purposeful implementation of energy. Thus, the only thing to now interfere with your living a healthier, and better, life is you . . .

## Useful Hebrew Words

| English | Hebrew (🔊) | A-B-Cs |
|---|---|---|
| Truth | Emet (I-m-et) | אמת |
| Life | Khayeem (Kaa-yem) | חיים |
| Love | Ahavah (A-ho-va) | אהבה |
| Peace | Shalom | שלום |
| Health | Breeoot (Bree-ou-oot) | בריאות |
| Thank You | Todah (Toe-da) | תודה |
| Victory | Neetsakhon (Nee-sa-hon) | ניצחון |

Furthermore, as for why I should dare to mention the Egyptians in a book that heralds the Anointed One of YHVH—the world's oldest Christian texts are not Greek or Latin—they are Egyptian! Scholars explain:

# Epilogue

*"Our earliest Christian manuscripts have much in common. All (with a single exception from the third century) come from Egypt..."*[1]

No less telling, the same is so of the world's first Christian monasteries. What you've been told—and what happens to be so—are often two entirely different matters. But until you're willing to know—you will not grow...

**Egyptian Coptic Monastery of Anba Hidra built in the 4th Century**

---

[1] Groves, C., The Planting of Christianity in Africa Vol. I, pp. 36 - 39 & Ackroyd, P., & Evans, C., The Cambridge History of the Bible Vol. I, pp. 28, 52 - 53, 56 – 57, 62, 601 & Monasticism, Encyclopedia of Early Christianity p. 615 & Pachomius, The Coptic Encyclopedia Vol. VI, pp. 1859 - 1860 & Jordan Maxwell: Egypt (DVD) & Maxwell, J., That Old Time Religion p. 38

The Christian terms *Amen* and *Hallelujah* were originally ancient Egyptian words; in truth, the latter actually meant, "Horus is Risen."

### YHSVH

**Yod-Hey-Shin-Vav-Hey**
*You lead me in the path of righteousness for your name's sake*

**Black Madonna painted in the 1600s by the Spanish painter Riccia. Catholics treasure it to this day!**

# Epilogue

As someone who has traveled the world and researched these matters for decades, I am quite comfortable sharing these truths with you; and rest assured, I am not the only one. For instance, Jordan Maxwell has declared: *"If you really wanted to get serious about Christianity, then I would suggest that you begin right here with Coptic [Egyptian] Christianity because that's where it all began . . . ."*[2]

But irrespective of how you view the Creator—know that the source of the electromagnetic energy that's pulsating within every cell of your body is the Divine![3] In conclusion, for so much as loving someone more than they love their self won't prove fruitful—I'm through here save this: *The victory is not contingent upon the enslavement of another. It comes after you liberate yourself . . .*

---

[2] Jordan Maxwell: Egypt (DVD)
[3] Lakhovsky, G., & Clement, M., Secret of Life: Cosmic Rays and Radiations of Living Beings  p. 75

# Endnotes

**USSL**

# DEPARTMENT OF DEFENSE APPROPRIATIONS FOR 1970

UNITED STATES SENATE LIBRARY

## HEARINGS
BEFORE A
## SUBCOMMITTEE OF THE COMMITTEE ON APPROPRIATIONS HOUSE OF REPRESENTATIVES
NINETY-FIRST CONGRESS
FIRST SESSION

---

SUBCOMMITTEE ON DEPARTMENT OF DEFENSE
GEORGE H. MAHON, Texas, *Chairman*

ROBERT L. F. SIKES, Florida
JAMIE L. WHITTEN, Mississippi
GEORGE W. ANDREWS, Alabama
DANIEL J. FLOOD, Pennsylvania
JOHN M. SLACK, West Virginia
JOSEPH P. ADDABBO, New York
FRANK E. EVANS, Colorado¹

GLENARD P. LIPSCOMB, California
WILLIAM E. MINSHALL, Ohio
JOHN J. RHODES, Arizona
GLENN R. DAVIS, Wisconsin

R. L. MICHAELS, RALPH PRESTON, JOHN GARRITY, PETER MURPHY, HUNTER SPILLAN,
ROBERT FOSTER, *Staff Assistants*

¹ Temporarily assigned

*H.B. 15090*

PART 5
RESEARCH, DEVELOPMENT, TEST, AND EVALUATION
Department of the Army
Statement of Director, Advanced Research Project Agency
Statement of Director, Defense Research and Engineering

---

Printed for the use of the Committee on Appropriations

U.S. GOVERNMENT PRINTING OFFICE
WASHINGTON : 1969

UNITED STATES SENATE LIBRARY

# Endnotes

## DEPARTMENT OF DEFENSE APPROPRIATIONS FOR 1970

### SYNTHETIC BIOLOGICAL AGENTS

There are two things about the biological agent field I would like to mention. One is the possibility of technological surprise. Molecular biology is a field that is advancing very rapidly and eminent biologists believe that within a period of 5 to 10 years it would be possible to produce a synthetic biological agent, an agent that does not naturally exist and for which no natural immunity could have been acquired.

Mr. SIKES. Are we doing any work in that field?

Dr. MACARTHUR. We are not.

Mr. SIKES. Why not? Lack of money or lack of interest?

Dr. MACARTHUR. Certainly not lack of interest.

Mr. SIKES. Would you provide for our records information on what would be required, what the advantages of such a program would be, the time and the cost involved?

Dr. MACARTHUR. We will be very happy to.

(The information follows:)

The dramatic progress being made in the field of molecular biology led us to investigate the relevance of this field of science to biological warfare. A small group of experts considered this matter and provided the following observations:

1. All biological agents up to the present time are representatives of naturally occurring disease, and are thus known by scientists throughout the world. They are easily available to qualified scientists for research, either for offensive or defensive purposes.

2. Within the next 5 to 10 years, it would probably be possible to make a new infective microorganism which could differ in certain important aspects from any known disease-causing organisms. Most important of these is that it might be refractory to the immunological and therapeutic processes upon which we depend to maintain our relative freedom from infectious disease.

3. A research program to explore the feasibility of this could be completed in approximately 5 years at a total cost of $10 million.

4. It would be very difficult to establish such a program. Molecular biology is a relatively new science. There are not many highly competent scientists in the field, almost all are in university laboratories, and they are generally adequately supported from sources other than DOD. However, it was considered possible to initiate an adequate program through the National Academy of Sciences-National Research Council (NAS-NRC).

The matter was discussed with the NAS-NRC, and tentative plans were made to initiate the program. However, decreasing funds in CB, growing criticism of the CB program, and our reluctance to involve the NAS NRC in such a controversial endeavor have led us to postpone it for the past 2 years.

It is a highly controversial issue and there are many who believe such research should not be undertaken lest it lead to yet another method of massive killing of large populations. On the other hand, without the sure scientific knowledge that such a weapon is possible, and an understanding of the ways it could be done, there is little that can be done to devise defensive measures. Should an enemy develop it there is little doubt that this is an important area of potential military (technological inferiority in which there is no adequate research program.

Source: Department of Defense Appropriations for 1970. Hearings Before a Subcommittee of the Committee on Appropriations House of Representatives, Ninety-First Congress, Tuesday, July 1, 1969, Page 129. Washington: U.S. Government Printing Office, 1969.

… YHVH …

# YHVH

### I

On dozens of occasions, the scriptures speak about the majesty of the Creator's holy name.  Maybe I am alone, but I don't think that YHVH gives us commands to hear Himself make noise.  If we are told to keep His name holy and our response is—*Sure, just as long as we can change it two or three times every thousand years!*  Well, we might not be dead wrong—<u>but we certainly ain't 100% right</u>!  Not only that, but the Anointed One also told us that we shall do greater things than He, if we pray to the Father in <u>His</u> name.[1]  Yet, how many of us *actually* have, or, *actually* do?

---

[1] <u>John</u> 13: 16 & <u>Hosea</u> 4: 6
Never forget, the one who sends is always greater than the one who is sent!  I can't help but be reminded of the passage in Hosea: *"My people are destroyed from lack of knowledge."*

## Endnotes

Moreover, if there isn't any significance to the actual name of the Creator—*Why have those who oppose Him done so much to hide and obscure His name from the people who have historically exhibited the most spirituality and reverence?* If it all has no significance, why should they feel the need to sink to these depths?

Look, according to the *Gospel of Thomas*: *"Whoever has known the World has found a corpse, and whoever has found a corpse, of him the World is not worthy."* The <u>infallible</u> word of the Creator is infallible precisely because <u>it establishes the celestial ordinances which control your destiny</u>. What needs to be understood is that, *intentions aside*, if you try to operate outside of the celestial ordinances, you are undertaking a precarious venture! Thus, the greatest source of spiritual sound energy for a Follower of the Way is activated by the words: *YHVH, in the name of YHSVH . . .*

יהשוה

# 𝔅𝔞𝔟𝔢𝔩

I

Constantine the Great, Encyclopedia of Early Christianity p. 326 & Constantine I, Universal Standard Encyclopedia Vol. VI, p. 1962 & Davis, E., The First Sex pp. 232 - 238 & Walker, B., The Woman's Encyclopedia of Myths and Secrets p. 820 & Bowder, D., Who was Who in the Roman World p. 106 & Frank, T., A History of Rome p. 562 & Constantine I, The New Columbia Encyclopedia p. 634 & Grant, M., The Roman Emperors pp. 227 - 234 & Boak, A., & Sinnigen, W., A History of Rome to A.D. 565 pp. 502 - 507 & Millar, F., The Emperor in the Roman World pp. 584 - 607 & Pfeffer, L., Church, State, And Freedom p. 14 & Du Bourguet, P., Early Christian Painting p. 10 & Harnack, A., The Mission and Expansion of Christianity in the first three centuries Vol. II, p. 75 & Hastings, J., Selbie, J., & Lambert, J., Dictionary of the Apostolic Church & Guillamont, A., The Gospel According to Thomas p. 19 & Kidd, B., Documents Illustrative of the History of the

### Endnotes

Church Vol. II p. 7 & Higgins, G., Anacalypsis Vol. I, p. 331, Vol. II, p. 51, 89 -90 & Hyde, W., Paganism to Christianity in the Roman Empire p. 257 & Exodus 31:12-18 & Ward, P., A Dictionary of Common Fallacies p. 249 & Veith, W., The Man Behind the Mask (DVD) & Horowitz, L., DNA: Pirates of the Sacred Spiral (DVD) & Hall, M., The Secret Teachings of All Ages p. CLVI & Directions, Man, Myth and Magic Vol. III, p. 640 & Muller, M., Mythology of All Races Vol. XII, p. 66 & Dampier-Whetham, W., A History Of Science: And its relations with Philosophy & Religion p. 5 & Tompkins, P., The Magic of Obelisks pp. 354 – 355, 433 & Mialon, E., The Great Pharaoh Ramses and His Time Ch. 2, 8 & De Lubicz, R., The Temple in Man p. 85 & Exodus 20:8-11 & Horowitz, L., & Puleo, J., Healing Codes for the Biological Apocalypse pp. 172 –175 & De Lubicz, R., Sacred Science: The King of Pharaonic Theocracy pp. 126 – 135, 159, 162 & Mark 7: 9 & Painting by William Hogarth appears courtesy of Directmedia

The Evil One has done everything in his power to flip proper things on their head: up is down; right is wrong, etc., etc. Little wonder for this when we reflect upon the fact that the word *Evil,* is literally the word *Live* turned backwards. Be well told, attempting to change a celestial mandate is serious business as it is a timeless decree—and yet:

> Despite being told to honor them—<u>the divine names have been changed so many times that they are not even known by most of us</u>—*never mind them actually passing our lips!*

> Second, despite the fact that we were told to honor the Sabbath (the 7th day of the week) it was changed in the 4th century CE. The Roman Emperor Constantine changed the Christian Sabbath from its original Saturday to Sunday to make the new faith more palatable to the citizens of the Roman State. In an edict entitled, *Constantine's Legislation about Sunday, 321 A.D.* we find the following:

## Endnotes

"*Constantine to Elpidius—All judges and city-people and the craftsmen shall rest upon the venerable Day of the Sun. Country-people, however, may freely attend to the cultivation of the fields, because it frequently happens that no other days are better adapted for planting the grain in furrows or the vines in trenches; so that the advantage given by heavenly providence may not, for the occasion of a short time, perish. Given on March 7 [321], in the second consulate of Crispus and the second of Constantine.*"

Evidently, YHVH's instruction to Moses, two-thousand years of Hebraic tradition, and almost three-hundred years of Christian custom—was less important to Constantine than continuing Rome's religious tradition of week day worship on the *Dies Solis* or the "Day of the Sun" (Sun-

Day). That notwithstanding, Thomas records the Christian Christ as saying: *"If you keep not the Sabbath as Sabbath, you will not see the Father."*

We even measure our days differently than as was first ordained. In ancient times days were measured like the structure of creation: first came the darkness (evening and night) then we have the light (morning and day time) and then the cycle repeats– *darkness then the light.*

However in the 16th century CE, Pope Gregory XIII changed the measure with the institution of the Gregorian Calendar. His rationale was that a change was necessary to bring the solar calendar year more in line with the true date of Easter. Gregory's papal bull called for the substitution of his new calendar for the Julian Calendar.

# Endnotes

The way days were actually measured would soon follow. By changing the measure by 6 hours (from 6 pm to 12 am), midnight became the beginning of the new day. Believe it or not, just as sound matters—so does time. I am reminded of the scripture: *"And he said to them: 'You have a fine way of setting aside the commands of God in order to observe your own traditions!'"*

**Famous portrait by William Hogarth c. 1755 CE. The scene is said to represent the English election to adopt the new Gregorian Calendar.**

𝔜𝔋𝔖𝔙𝔋

# 𝔅𝔞𝔟𝔢𝔩

## II

Gordon, C., Before the Bible pp. 27 - 28, 91 & Malcioln, J., The African Origins of Modern Judaism pp. 3 - 4 & Murphy, E., Diodorus on Egypt p. 35 & Cox, G., African Empires and Civilizations p. 14, 45 & Burn, A., & Selincourt, A., Herodotus: The Histories p. 114, 121 & Jackson, J., An Introduction to African Civilizations p. 66, 71 & Wright, G., Biblical Archeology pp. 36 - 37, 42 - 43 & Great Soviet Encyclopedia Vol. XI, p. 481, Vol. XXVIII, p. 530 & Gray, J., Near Eastern Mythology pp. 8, 14, 20, 22, 25, 70 - 73 & Ancient Mesopotamia pp. 40 - 41, 59 & Gurney, J., Kingdoms of Asia the Middle East and Africa pp. 144 - 145 & Jobes, G., Dictionary of Mythology, Folklore and Symbols p. 1065 & Windsor, R., From Babylon to Timbuktu pp. 15, 24, 27 -28 & Saggs, H., Civilization before Greece and Rome pp. 8, 17 - 18, 37, 40 - 41, 156 - 162 & Kephart, C., Races of Mankind pp. 51 - 53, 118 - 120, 146, 250, 252, 256, 268 - 269, 348 & Illustrated Dictionary & Concordance of the Bible pp. 23, 107 - 108, 304,

450, 454 - 455, 500 - 501, 724 & Maspero, G., History of Egypt Vol. VI, p. 158, Vol. XIII, p. 289 & Diggs, E., Black Chronology p. 2 & Rogers, J.A., Sex and Race Vol. I, pp. 58 - 59, 110, 266 & Grant, M., The History of Ancient Israel p. 13 & Oates, J., Babylon pp. 24, 32 - 35, 58, 86 - 89, 95 - 97, 104 - 111, 170 – 174, 199 - 202 & Elam, The Encyclopedia Britannica Vol. IV, p. 416, Vol. VI, pp. 758 - 759, Vol. XXIV, p. 111 & Brinkman, J., Materials and Studies for Kassite History p. 8, 29 & Noblecourt, C., Tutankhamen: Life and Death of a Pharaoh p. 213, 291 & Cotterell, A., The Penguin Encyclopedia of Ancient Civilizations pp. 72 - 73, 93 & Sykes, E., Everyman's Dictionary of Non-classical Mythology p. 71, 104 & Kramer, S., From the Tablets of Sumer & Fairservis, W., Mesopotamia: The Civilization that Rose out of Clay pp. 67 - 71 & Elam, The Columbia Encyclopedia p. 556 & Brunson, J., Kamite Brotherhood: African Origins in Early Asia pp. 5, 31, 39 - 40 & Hicks, J., The Persians & Collins, R., The Medes and Persians & The Cambridge Ancient History Volume I Part II: Early History of the Middle East pp. 644 - 645, 647, 651, 680 & Higgins, G., Anacalypsis Vol. I, p. 264, 316, 399, 462, Vol. II, pp. 137 – 139 & Davis, E., The First Sex p. 123 & Hinz, W., The Lost World of Elam & Encyclopedia of Religion Vol. I, pp. 367 - 368, Vol. II, pp. 162 - 163, Vol. VII,

𐤉𐤄𐤔𐤅𐤄

pp. 216 - 271, Vol. IX, pp. 449 - 451 & Elam, <u>New Columbia Encyclopedia</u> p. 845 & Hodgson, M., <u>The Venture of Islam: Conscience and History in a World Civilization</u> Vol. I, p. 142 & Elamite, <u>Collier's Encyclopedia</u> Vol. VIII, p. 685 & Eban, A., <u>Heritage: Civilization and the Jews</u> p. 14, 37 & Mac Culloch, C., <u>Mythology of All Races</u> Vol. V, p. 11 & Whiston, W., <u>The Life and Works of Flavius Josephus</u> p. 41 & Carlyon, R., <u>Guide to the Gods</u> p. 299 & Jordan, M., <u>Encyclopedia of Gods</u> p. 92 & Dalman, G., <u>Jesus - Jeshua</u> p. 16 & Ackroyd, P., & Evans, C., <u>The Cambridge History of the Bible</u> Vol. I, pp. 1, 5 - 6, 13 & <u>Genesis</u> 36: 9

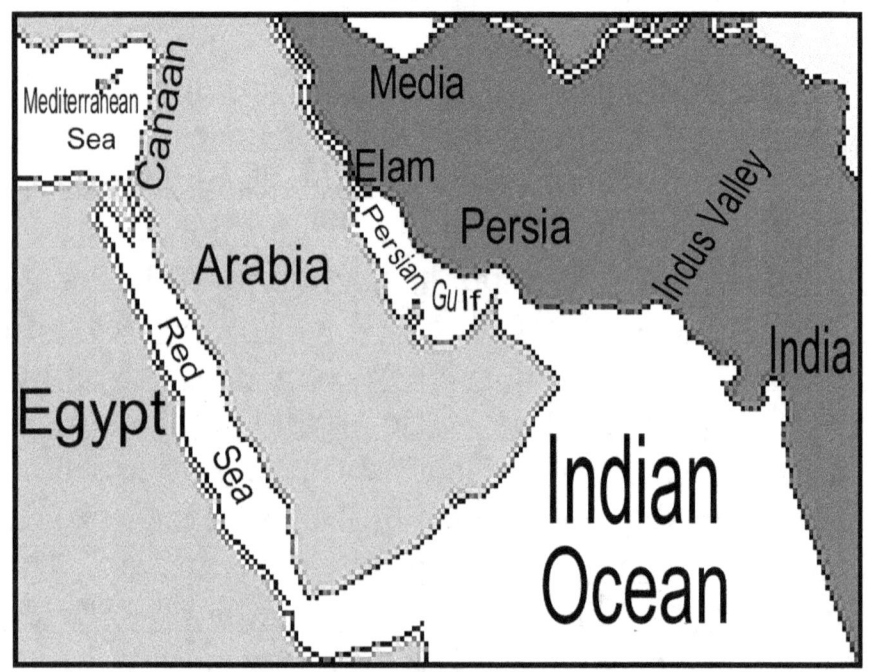

**Map of Southwestern Asia**

## Endnotes

Because there is so much confusion over the racial identity of the first peoples of the Fertile Crescent, I have taken the liberty to not only supply you with the references I've used in this work, but several other useful works. It is my hope that this will give you a chance to obtain a clear understanding of this matter should you so choose.

Anthropologists place Hametic populations in ancient Persia (modern Iran) as early as 8000 BC. Artifacts of the ancient Medes and Persians clearly substantiate this finding (see page 83 and 110). The Elamites called their nation *Hal-ta-mi*, which means "Land of the Lord." The capital city of Elam was Susa; hence, the reason Elam is also referred to as Susiana. Elam was bordered by Assyria and Madai (Media) in the north—the Persian Gulf (which in their day was 100 miles longer than it is currently) in the south—and the Tigris River in the west.

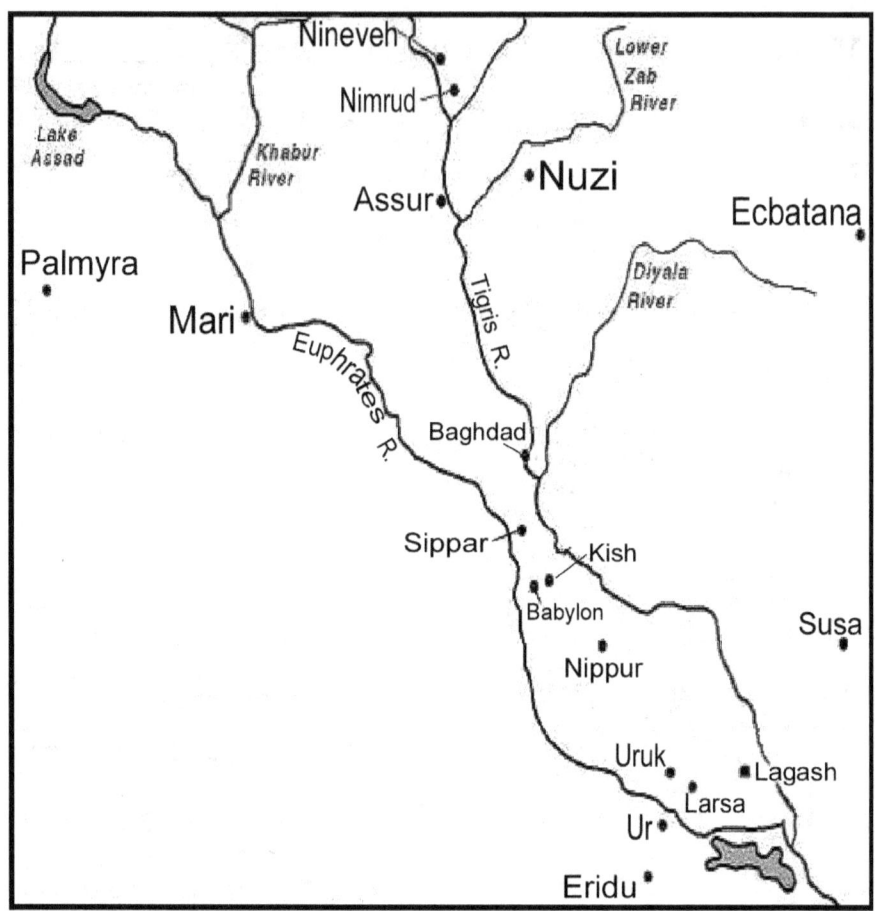
**Ancient region of the Tigris and Euphrates Rivers**

Scholars explain that a dynasty of no less than three Elamite kings ruled Ur during the 27th century BC. However, the northerly invasion of the Elamites into Mesopotamia during the 3rd millennium BC was ultimately squashed by the Amorites (Hamites). Quoting Maspero, *"The*

## Endnotes

*Elamites had never abandoned their efforts to press in every conceivable way their claim to the supremacy, which, prior to Hammurabi, had been exercised by their ancestors over the whole of Mesopotamia..."*

Elam's 12th century BC decline in Mesopotamia came when Babylon's Nebuchadnezzar I (reigning c.1124 BC - 1103 BC) mounted a great revolution culminating in the sacking of Susa. Incidentally, even though the language and pictographs of the proto-Elamites (an agricultural people who had probably settled in the region by 7000 BC) were different from the early Mesopotamians, by the 1st millennium BC the Elamites were proficient in the Akkadian cuneiform script of the Babylonians.

To continue on, Kephart speaks of Hamites dwelling in the Persian region clear down into modern times! It is noteworthy that the religion of the Arameans also fell under the spiritual umbrella of the Canaanites. One of the greatest deities of the ancient Canaanites was called *Adad* (at times pronounced *Hadad*). Unsurprisingly, he was a powerful ruler of the atmosphere, thunder and rain. Additionally, the alphabet of the Edomites of Canaan (modern day Southern Israel) was also a Canaanite script. In passing, the Edomites were descendants of Esau (the brother of Jacob of the Hebrews).

**Bust of an ancient Canaanite Man**

# Endnotes

# Under the Sun

## I

Alchemy, The American Peoples Encyclopedia Vol. I, p. 537 & The Kybalion: Hermetic Philosophy p. 44 & DeGivry, G., Witchcraft, Magic & Alchemy & Hentz, F., & Long, G., Experiments with Chemical Reactions & Element, Chemical, The World Book Encyclopedia Vol. V, pp. 2280 - 2283 & Lucas, A., Ancient Egyptian Materials & Industry & Sadoul, J., Alchemists and Gold & Hall, M., The Secret Teachings of All Ages pp. XXIV, LIII - LIV, CLIX - CLX & Alchemy, Universal Standard Encyclopedia Vol. I, p. 142 & Higgins, G., Anacalypsis Vol. II, p. 398, 543 & Wilson, J., Culture of Ancient Egypt p. 59 & (Audio), Ancient African Medical Practices C. Finch 1992 & Donnelly, I., & Sykes, E., Atlantis: The Antediluvian World p. 188 & De Lubicz, R., Sacred Science: The King of Pharaonic Theocracy pp. 21- 22, 24 - 26, 38, 98, 114, 178, 258 - 260, 274, Apx. IV & Budge, E.A., The Dwellers on the Nile p. 193 & Burn, A., & Selincourt, A., Herodotus: The Histories p. 130, 160, 169, 191, 195, 200 & Smith, W., Dictionary of Greek and Roman Biography and Mythology Vol. II, pp. 1107 - 1108,

Vol. III, p. 616 & Murphy, E., Diodorus on Egypt pp. 13, 90 - 91, 104, 105, 129 - 130 & Saunders, J., The Transitions From Ancient Egyptian to Greek Medicine pp. 8 – 9, 14 – 15, 30 & Sedgwick, W., & Tyler, H., A Short History of Science pp. 37 – 38 & Sedgewick, W., Tyler, H., & Bigelow, R., A Short History of Science p. 36 & Isocrates, Busiris Norlin, G., tr. (net) & Rhind Papyrus, The Encyclopedia Britannica Vol. X, p. 21 & Papyrus, & Cyrene, The Great Soviet Encyclopedia & Cyrenaica, The Universal Standard Encyclopedia Vol. VI, pp. 2186 - 2187 & Historical Atlas of Africa p. 16 & Cyrene, The American Peoples Encyclopedia Vol. VI, p. 690 & Tompkins, P., The Secret of the Great Pyramid pp. 112 - 113 & Rogers, J.A., Sex and Race Vol. I, p. 88 & Fraser, P., Eratosthenes of Cyrene pp. 4 - 5, 22 - 26, 30 - 31, 33 & Snowden, F., Blacks in Antiquity p. 170 & Gordon, C., Before the Bible pp. 115 – 116 & Waddell, W., Manetho p. 33 & David, R., The Egyptian Kingdoms p. 115 & Dampier-Whetham, W., A History Of Science: And its relations with Philosophy & Religion p. 7 & Grossinger, R., Planet Medicine p. 314 & Casson, L., Daily Life In Ancient Egypt p. 64 & Budge, E.A., The Dwellers on the Nile pp. 188, 196 - 198 & Thornwald, J., Science and Secrets of Early Medicine p. 54 & McGrew, R., Encyclopedia of Medical History pp.

# Endnotes

47 - 48, 321 & Maspero, G., <u>History of Egypt</u> Vol. Vl, p. 70 & Thomson, W., <u>Medicine From the Earth: A Guide to Healing Plants</u> p. 10 & Iakovids, S., <u>Mycenae - Epidaurus: Argos - Tiryns - Nauplion</u> p. 130 & ben-Jochannan, Y., <u>Black Man of the Nile and His Family</u> pp. 254 - 255, 324 - 331 & Bowder, D., <u>Who was Who in the Greek World</u> pp. 53, 67, 72 - 73, 101, 104, 107, 148 & Stein, W., <u>Atlantis</u> & Muck, O., <u>The Secret of Atlantis</u> pp. 12 - 13 & Stemman, R., <u>Atlantis and Lost Lands</u> p. 90 & Cavendish, M., <u>Man, Myth & Magic</u> & Berlitz, C., <u>Atlantis: The Eighth Continent</u> & Sjoo, M., & Mor, B., <u>The Great Cosmic Mother</u> p. 22, 32 & Ghalioungui, P., <u>Magic and Medical Science in ancient Egypt</u> p. 152 & Cox, G., <u>African Empires and Civilizations</u> pp. 40 – 43 & Diop, C.A., <u>The African Origin Of Civilization: Myth Or Reality</u> p. 232 & Kedourie, E., <u>Nationalism in Asia and Africa</u> p. 278 & Malcioln, J., <u>The African Origins of Modern Judaism</u> p. 225

There isn't any doubt that the Greeks were beneficiaries of the knowledge of the ancient Egyptians. This is especially so in regards to the disciplines of chemistry, astrology, medicine and mathematics! While I won't be voluble, permit me to share the following:

The Egyptians possessed an amazing comprehension of the manipulation of the elements; e.g., using mercury to purify metals before 1500 BC. Indeed, their understanding is even believed to have extended into the craft of <u>Alchemy</u> (the science of spontaneously transmuting lead into gold).

That the Greeks were amazed by the Egyptians' comprehension of the elements is manifest. For example, the ancient Egyptians long held that the God <u>Atum</u> was the all-encompassing and omnipotent creator: <u>the All within Himself</u>! One other way to express this idea was *"Atum was in all things!"* Lo and behold—after attending schools of science and philosophy for five

## Endnotes

years in Egypt during the 5th century BC—the acclaimed Greek philosopher Democritus would return to Greece with the brilliant idea that everything is made up of little Atums (today spelled <u>Atoms</u>). Further here, Strabo would make the disclosure that a Phoenician (Canaanite) scholar of Sidon named Mochus (who lived c.1300 BC) was to actually author a work on atomic theory centuries before Democritus was born.

Ancient Greek writers also maintained that the Egyptians were the first people to practice astrology. Diodorus explained: *"Oenopides was in like manner a disciple of the priests and astrologers and learned many things, especially that the sun's orbit is an oblique course and traces a retrograde path opposite to that of the other stars. In somewhat the same way*

*Eudoxos studied astronomy among them [the Egyptians] and attained an eminent reputation by transmitting much useful knowledge to the Greeks."*

The historian would go on to say of the Egyptians: *"They also carry geometry and arithmetic to the highest perfection . . ."* By the 2nd millennium BC, the Egyptians were not only calculating the volume of truncated pyramids with square bases—but also determining the areas of curved surfaces! When one stops to reflect upon the fact that as late as the 9th century BC the Hellenes (Greeks) had not begun to utilize an alphabet or a system of numbers—there can be no doubt as to the course of the diffusion of ancient scientific and philosophical understanding.

# Endnotes

As a matter of fact, the Greek historian Plutarch went so far as to name the Heliopolian Priest <u>Oenuphis</u> as being Pythagoras's instructor. You remember Pythagoras and the celebrated <u>Pythagorean Theorem</u>: the square on the hypotenuse side of a right triangle is equal to the sum of the squares of the other two sides—and all is harmony. To quote Socrates:

*"If one were not determined to make haste, one might cite many admirable instances of the piety of the Egyptians, that piety which I am neither the first nor the only one to have observed; on the contrary, many contemporaries and predecessors have remarked of it, of whom Pythagoras of Samos is one. On a visit to Egypt he became a student of the religion of the people, and was first to bring to the Greeks all philosophy, and more conspicuously than others he*

*seriously interested himself in sacrifices and in ceremonial purity, since he believed that even if he should gain thereby no greater reward from the gods, among men, at any rate, his reputation would be greatly enhanced."*

Western scholars commonly accredit Euclid, author of *The Elements*, with creating the science of geometry during the 4th century BC. But as has hitherto discussed, the Egyptians had created and were employing the science centuries before the birth of Euclid. What's more, as the mathematician is never reported as living in any land other than Egypt—it is curious that he should be regarded as "Greek." Yet, for some definitive testimony from the ancients about the matter, allow me to share this observation by Herodotus:

## Endnotes

*"Any man whose holding was damaged by the encroachment of the river would go and declare his loss before the king, who would send inspectors to measure the extent of the loss, in order that he might pay in future a fair proportion of the tax at which his property had been assessed. Perhaps this was the way in which geometry was invented, and passed afterwards into Greece..."*

In truth, the Ionians and many of the most revered figures in Hellenic history were educated in Egypt. Diodorus remarked:
*"For many of the ancient customs that were current among the Egyptians were valued not only among the native inhabitants, but were also greatly admired by the Greeks. For this reason, Greeks of the highest repute for learning were eager to visit Egypt, that they might*

*gain knowledge of its noteworthy laws and customs . . ."*

For so much as it was common knowledge throughout the Old World that Orpheus, Musaeus, Archimedes, Hecataeus, Homer, Lycurgos, Solon, Pythagorus, Plato, Eudoxos, Democritus, and Oenopides were all trained by Egyptians—allow me to close with these words by Amelineau:

*"I then saw, and saw clearly, that the most famous systems of Greece, notably those of Plato and Aristotle, had Egypt for a cradle. I also realized that the Greeks' fine genius, particularly that of Plato, had been able to clothe Egyptian ideas with an incomparable dress; but I thought that what we loved when coming from the Greeks, should not be disdained or looked down upon when found among the Egyptians. When, in our own day, two*

## Endnotes

*authors collaborate, they share the fame which accrues from their work: I do not see why ancient Greece should keep for herself alone the honour due to ideas which she had borrowed from Egypt."*

**Egypt's Hall of Columns leading to Obelisk at Thebes**

# YHSVH

# Under the Sun

## II

II Corinthians 4:18 & Murphy, E., Diodorus on Egypt pp. 13, 90 - 91, 104, 105 - 106, 129 - 130 & Saunders, J., The Transitions From Ancient Egyptian to Greek Medicine pp. 8 – 9, 14 – 15, 30 & Dampier-Whetham, W., A History Of Science: And its relations with Philosophy & Religion pp. 2 – 7 & Sedgewick, W., Tyler, H., & Bigelow, R., A Short History of Science p. 31 - 38 & Gouldner, A., The Hellenic World p. 116 & Casson, L., Daily Life In Ancient Egypt p. 60 & Burn & Selincourt., Herodotus: The Histories pp. 158, 210 - 211 & Hippocratic Oath, Universal Standard Encyclopedia Vol. XII, pp. 4339 - 4440 & ben-Jochannan, Y., Black Man of the Nile and His Family p. 255 & Hall, M., The Secret Teachings of All Ages p. XXXVII & Tompkins, P., The Magic of Obelisks p. 47 & Ghalioungui, P., Magic and Medical Science in ancient Egypt pp. 68 – 71, 149 & Nova: Pyramids: Interview with Dr. Zahi Hawass, Director of the Pyramids (Net) & Camp, J., The Healer's Art: The Doctor through History pp. 26 - 27 & Ancient African Medical Practices C. Finch 1992 & David, R., The Egyptian Kingdoms p. 114 & Estes,

## Endnotes

J., <u>The Medical Skills of Ancient Egypt</u> p. 24 & Maspero, G., <u>History of Egypt</u> Vol. IX, p. 259 & Ancient Egypt Fan: <u>Ancient Egyptian Medical Papyri</u> (Net) & Bishop, M., <u>The Horizon Book of the Middle Ages</u> pp. 238, 244 - 246 & Cartwright, F., <u>Disease and History</u> pp. 54, 58 - 59 & Bloodletting, <u>Encyclopedia of Medical History</u> pp. 33 - 34 & Medicine, <u>The American Peoples Encyclopedia</u> Vol. XIII, p. 374 & Rogers, J.A., <u>Sex and Race</u> Vol. II, p. 398 & Zahoor, A., <u>Hospitals and Medical Schools in the Dark and Middle Ages</u> (net) & Van Sertima, I., <u>The Golden Age of the Moor</u> pp. 201, 210 - 211, 218 - 220, 226 - 230, 391 - 393 & Lacroix, P., <u>Science and Literature in the Middle Ages and at the Period of the Renaissance</u> pp. 145 – 146, 159 - 160 & Evans, J., <u>Life in Medieval France</u> p. 5 & Chejne, A., <u>Muslim Spain</u> pp. 347, 350 - 358 & Kristeller, P., <u>Renaissance Thought and Its Sources</u> pp. 120 - 121 & Bishop, M., <u>The Horizon Book of the Middle Ages</u> p. 246 & Aviccena, <u>Cambridge Dictionary of Philosophy</u> p. 56 & Avicenna, <u>McGraw-Hill Encyclopedia of World Biography</u> Vol. I, pp. 305 - 306 & Avicenna, <u>Encyclopedia of Philosophy</u> Vol. I, pp. 226 - 229 & Avicenna, <u>Webster's Biographical Dictionary</u> p. 80 & Ibn Zuhr, <u>Dictionary of Scientific Biography</u> Vol. XIV, p. 637 & Rashdall, H., <u>The Universities of Europe in the Middle Ages</u>

## YESUE

Vol. II, pp. 118 - 120, 127 – 128, 135 - 136 & HarperCollins Encyclopedia of Catholicism p. 1026

In addressing the subject of medicine, even though Western writers routinely declare Hippocrates to be the *Father of Medicine*—the fact is Imhotep (and many other Egyptians) practiced empirical medicine millenniums before there was a Greek State! Moreover, <u>it is rather apparent that the Greeks understood the origins of the science as they went so far as to proclaim Egypt's Imhotep the *God of Medicine*</u>!

Insomuch as the ancient Egyptians possessed medical texts on the human anatomy, created many effective medicates from plants, established specialists to treat specific illnesses, sewed wounds, trained nurses and even performed surgery before the Greeks had an alphabet—is it any wonder that Whetham should explain:

# Endnotes

*"Indeed, Egyptian medicine generally had an extensive influence. It spread to Greece, perhaps by way of Crete, and from Greece and Alexandria, in after ages, it passed into Western Europe..."*

**Imhotep reading a medical papyrus c. 2700 BC. This was 2300 years before Hippocrates was born. The Greeks worshipped this man as Asclepius—God of Medicine. The truth is that the genesis of many practices in Western medicine are, in fact, Egyptian.**

Actually, it would not be a stretch to characterize the Egyptians as the healthiest people in the Old World. In deference here to the ancient Greek historian Diodorus: *"The whole manner of life of the Egyptians was so evenly ordered that it would appear as though it had been arranged according to the rules of health by a learned physician rather than by a law-giver."*

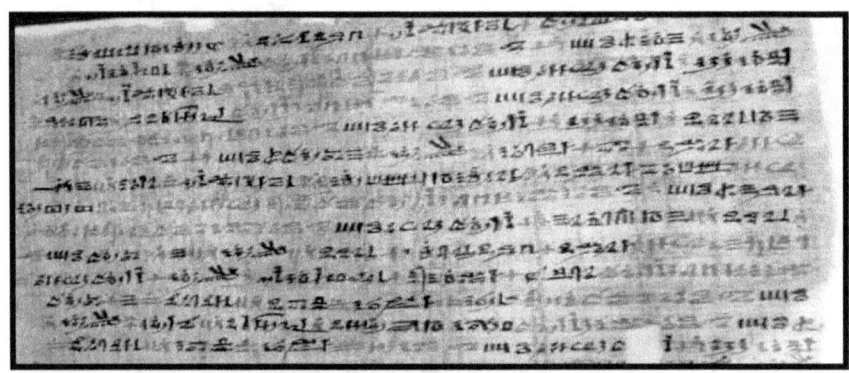

**Portion of an ancient Egyptian medical papyrus c. 1150 BC. This papyrus was originally owned by the Egyptian scribe Qen-her-khepeshef who lived during the 19th Dynasty. In complete form, the scroll discusses more than 40 illnesses.**

Additionally, the personal names and greetings of the Egyptians reflected their regard for good health. The Egyptian word for health was *Seneb*, and we find it in many of their names and

## Endnotes

salutations; e.g., phrases meaning, *"I possess health"* and *"May your father be healthy."*

We also learn that the practice of medicine in Egypt was much more a calling than a way to *make a living*. Accordingly, medical wages and fees were officially outlined and paid by the Egyptian government. **In this way, a doctor's medical decision-making and material gain were never connected to, or contingent upon, the suffering of others**.

That the Greeks were awestruck by Egypt's physicians is completely understandable. You see, throughout their history many Egyptians lived well into their fifties, sixties, seventies and beyond! *If an Egyptian noble died in their forties, somebody wasn't doing their job.* By contrast, in the 8th century BC, the life span of most Europeans was only about thirty-five years. As incredible as it may sound, I must share this overall assessment of Egyptian medicine by Camp: *"With their ideas on drugs, on diet, and on a free health service for the poor, the doctors of Egypt and Mesopotamia seem to be much nearer our own time than do the barber-surgeons of the Middle Ages."*

The reason Camp refers to surgeons of Medieval Europe as *"barbers"* is because thousands of years after the life of Imhotep—their prerequisite for becoming a doctor was simply knowing how to cut and shave hair! Incredibly, the following practices were commonplace amongst the physicians of Europe: if a person's limb became infected or fractured, the standard procedure was to cut the appendage off; peasants are known to have been given the boiled fat of felons as a cure for ails; and blood letting (the draining of random amounts of blood at various times from the body) was thought to cure everything from inflammation to infection.

## Endnotes

In truth, Western European medicine would not begin to advance from this state before the medical practices of the Moors began to spread in Europe. Moorish cities had hospitals that were open 24-hours a day to all comers regardless of the patient's ability to pay. These hygienic medical centers contained baths, food, and designated wings for specific ailments!

These sites were also staffed by salaried doctors and nurses who were trained in the use of drugs (created from herbs), cauterization, and surgery. It should further be noted that these physicians are known to have made associations between pleasant environs and a positive outlook with the acceleration of the healing process. Of the Moors' medical academy at Montpellier, Rashdall writes: *"No Medieval physician stood higher than Arnald of Villanova, Bernald de Gordon, and the other Montpellier doctors of the period . . ."*

Two Moors playing chess c.1270 CE. The woman on the right is playing a harp.

# Endnotes

# Under the Sun

## III

Goodman, F., Magic Symbols pp. 16 – 17 & Ward, P., A Dictionary of Common Fallacies p. 2 & Quirke, S., Ancient Egyptian Religion p. 26 & Hermes Mercurius Trismegistus: his Divine Pymander in Seventeen Books Bk. III & Hall, M., The Secret Teachings of All Ages p. XCIII & Tompkins, P., The Magic of Obelisks pp. 350 - 353 & Meyer, M., The Gnostic Gospels of Jesus & Oates, J., Babylon p. 198 & Schulberg, L., Historic India & (VHS), Transformation of Myth Through Time: From Psychology to Spirituality W. Free 1989 & Bunson, M., A Dictionary of Ancient Egypt p. 86 & John 2:19-22 & Worthy, R., About Black Hair & Image credit: U.S. Department of Energy Human Genome Program (net) & Naked Truth (VHS) I.R.E.S. 1991 & Akbar, N., Chains and Images of Psychological Slavery p. 22

## YESUE

Though aware of the fact that Dr. Horowitz is not an historian, I must say that his exaltation of the *Sacred Spiral* is not without historical foundation. You see, the circle was a very profound symbol to the ancients: for example, (1) the Egyptian God Thoth spoke of the circular operations of the neters; (2) the Mansion of Eternity was thought to be round; (3) some spiritual dances were performed in the round; (4) the Egyptians understood that the earth was round and that planets moved in a circular motion; (5) indeed Time itself was said to function as a circle; and (6), the circle was representative of beginning without end.

In addition, we find that this reverence for the circle was present in other parts of the Old World. For instance, the circle was a centerpiece of ancient Babylonian art. The *Chakra* of Buddhists (symbolic of the universal life system) is depicted as a circle or wheel. Coming down to

## Endnotes

these times, we even find modern scientists theorizing that the course of the universe is cyclical. To quote Goodman: *"Without doubt, the circle is the most important of all units in magical symbolism, and in almost every case where it is used, the circle is intended to denote spirit, or spiritual forces."*

**Celestial zodiac memorialized in stone c. 150 BC
Note the water bearer of the *Book of Luke***

But topping all of this (as Horowitz highlights) biologists studying the human genome explain that the spherical pattern is found when examining the essence of human life—DNA molecules being so structured as to form a double helix (or two spirals). As noted, the pattern also omnipresent present in Black hair.

This is a tiny snippet from the end of a lock of Black hair. When the hair is not damaged, it grows in a circle and retains that shape. It is the gathering of the individual circles (along with the activated melanin in the hair) that facilitates the hair forming into columns of wool or *locks*. In addition, *as previously stated*, this melanin also absorbs electromagnetic energy and nutrition from the ultraviolet rays of the sun!

# Endnotes

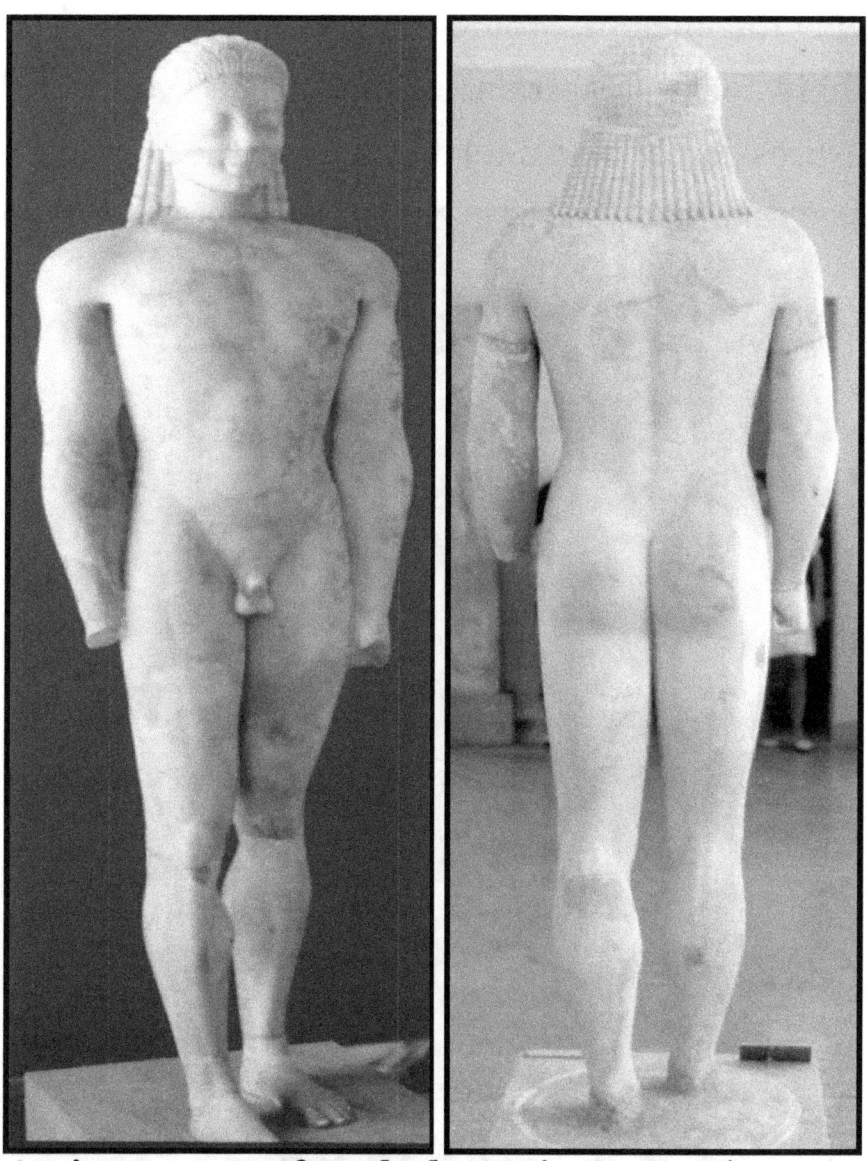

**Ancient statues of two lock wearing Canaanite men.**

# YHSVH

This understanding makes it rather easy to see why the ancients would embrace, *and not reject*, their locking spirals. They are a wonderful expression of man's relationship to the creation.

**King Seti I c. 1300 BC. The cobra was called Uraeus. It was an agent of righteousness to protect the Egyptians against their enemies.**

# Endnotes

As for why many Blacks today are not comfortable growing long and healthy locked hair—the unfortunate answer lies in self-ignorance and racism. When one stops to reflect upon the fact that Blacks have to expend great sums of money, invest a substantial amount of time, and even subject themselves to health risks to impede the natural growth of their hair—there isn't any other conclusion that one can come to.

Since no one has encapsulated the matter better, allow me to defer here to the respected Black clinical psychologist Na'im Akbar:

> *"This sense of inferiority still affects us in many ways. Our inability to respect African-American leadership, our persistent and futile efforts to look like and act like Caucasian people, is based upon this sense of inferiority. The persistent tendency to think of dark skin as unattractive, kinky [or locking] hair as*

'bad' hair, and African features as less appealing than Caucasian features, comes from this sense of inferiority. The lack of respect for African-American expertise and the irresponsibility of many African-American experts comes from this sense of inferiority. The disastrously high Black-on-Black homicide rate is in many ways indicative of fundamental disrespect for Black life growing out of this same sense of inferiority. It is a simple fact that people who love themselves seek to preserve their lives—not destroy them."

**The ancient Ankh was a very prominent Black symbol for life in the Old (or First) World**

## Endnotes

In a very real sense, Akbar hits the nail right on the head so-to-speak. What this self-rejection, at least on the subconscious level, says is this:

> *I understand the program of disdain for Black people. In fact, I understand it so well that I don't feel a need to respect myself! Indeed, I will undertake any futile act I can to frustrate my African nature and identity—starting with my hair . . .*

No less grave, Akbar is further substantiated by the fact that a blind adherence to a death culture's corporate media manipulation—has turned one generation's *Man, Brother, Queen, Lady, Sister, Afro* and *Harambee* conscious energy into another's *Nigger, Dog, Bitch, Ho, Shorty, Nappy* and to be *"In the life"* you got to be *"Bringing the Death"*—state. Try as one may, there isn't any way to overstate the significance of all this. **You know what I'm saying—and if you don't—you'd better 'axk' somebody!**

# YHSVH

Fortunately, our ancestors did not suffer these maladies. Wonder how I know—*it shows* . . .

**Seti I and the Goddess Hathor c. 1300 BC**

# Bibliography

Ackroyd, P., & Evans, C., The Cambridge History of the Bible  Cambridge Univ. Press 1970

Adachi, K., Rexresearch.com: The Story of Blood Electrification  www.rexresearch.com/kaali/kaali.htm 2007

Akbar, N., Chains and Images of Psychological Slavery  New Mind Productions 1984

Arnett, W., The Predynastic Origin of Egyptian Hieroglyphs  University Press 1982

Bailey, D., & Wright, E., Practical Fiber Optics  Newes 2003

Baines, J., & Malek J., Atlas of Ancient Egypt  Facts on File 1980

Bauval, R., & Gilbert, A., The Orion Mystery  Crown Pub. 1994

Becker, R., & Selden, G., The Body Electric: Electromagnetism And The Foundation of Life  Morrow 1985

ben-Jochannon, Y., Black Man of the Nile and His Family  African Heritage 1972

Berlitz, C., Atlantis: The Eighth Continent  Putnam & Sons 1984

Berlitz, C., Mysteries From Forgotten Worlds  Doubleday 1972

Berlitz, C., The Bermuda Triangle  Doubleday 1974

Bishop, M., The Horizon Book of the Middle Ages  American Heritage & Bonaza Books 1968

Boak, A., & Sinnigen, W., A History of Rome to A.D. 565  Macmillan 1965

Bowder, D., Who was Who in the Roman World  Cornell Univ. Press 1975

# Bibliography

Bowder, D., Who's Who in the Greek World  Phaidon 1982

Brander, B., The River Nile  National Geographic 1966

Breasted, J., A History of Egypt: From the Earliest Times to the Persian Conquest  C. Scribner's Sons 1909

Brinkman, J., Materials and Studies for Kassite History Vol. I  Oriental Institute 1976

Brunson, J., Kamite Brotherhood: African Origins in Early Asia  Brunson & Kara Publ. 1989

Bucaille, M., Mummies of the Pharaohs  St. Martin's Press 1990

Budge, E.A., A History of Egypt  Anthropological Pub. 1968

Budge, E.A., A Short History of the Egyptian People  Dent 1923

Budge, E.A., An Egyptian Hieroglyphic Dictionary  Dover 1978

Budge, E.A., Egypt  Holt 1925

Budge, E.A., Egyptian Language  Routledge & Paul 1963

Budge, E.A., Egyptian Magic  Routledge & Paul 1963

Budge, E.A., Egyptian Religion  Dell Press 1959

Budge, E.A., Mummy: A Handbook of Egyptian Funerary Archaeology  Kessinger Publ. 2003

Budge, E.A., Osiris: & The Egyptian Religion of Resurrection Vol. I - II  Univ. Books 1961

Budge, E.A., The Book of the Dead  Carol Publ. 1994

Budge, E.A., The Dwellers on the Nile  Dover 1977

Bunson, M., The Encyclopedia of Ancient Egypt  Facts on File 1991

Burn, A., & Selincourt, A., Herodotus the Histories Penguin 1972

Butzer, C., (Ed.)., Ancient Egypt: Discovering Its Splendors National Geographic Soc. 1978

Calladine, C., & Drew, H., Understanding DNA: The Molecule & How IT Works  Accademic Press 1966

Camp, J., The Healer's Art: The Doctor through History Taplinger 1977

Candille, S., What Makes Us Human?  The Tech Museum/Understanding Genetics http://www.thetech.org/genetics/news.php?id=8  2007

Carlyon, R., A Guide to the Gods  Morrow & Co. 1982

Carlyon, R., Guide to the Gods  Heinmann 1981

Cartwright, F., Disease and History  Crowell 1972

Casson, L., Daily Life in Ancient Egypt  Time 1972

Cavendish, M., Man, Myth & Magic  B.P.C. Publ. 1970 - 1995

Champion, S., Racial Proverbs  Barnes & Noble 1938

Chejne, A., Muslim Spain  Univ. of Minn. 1974

Clark, J., Yallop, C., & Fletcher, J., An Introduction to Phonetics and Phonology  Blackwell 2006

Clark, R., The Sacred Tradition in Ancient Egypt: The Estoteric Wisdom Revealed  Llewellyn 2000

Clayman, C., The Human Body: An Illustrated Guide to its Structure, Functions, and Disorders  Dorling Kendersley 1995

Collins, R., Medes and Persians: Conquers & Diplomats McGraw-Hill 1972

Cotterell, A., The Penguin Encyclopedia of Ancient Civilizations  Penguin 1980

# Bibliography

Cox, G., African Empires and Civilizations  African Heritage Studies Publ. 1974

Crowley, B., Words of Power: sacred sounds of East & West  Llewellyn Publ. 1991

Dalman, G., Jesus - Jeshua  Macmillan 1929

Dampier-Whetham, W., A History Of Science: And its relations with Philosophy & Religion  Macmillan 1931, Cambridge Univ. 1948

Daniel, G., Nubia Under the Pharaohs  Westview Press 1976

David, R., The Egyptian Kingdoms  Elsevier & Phaidon 1975

Davies, W., Egyptian Hieroglyphs  Univ. of Calif. Press 1987

Davis, E., The First Sex  Putnam & Sons 1971

De Lubicz, R., Sacred Science: The King of Pharaonic Theocracy  Inner Traditions International 1961

De Lubicz, R., The Temple in Man  Inner Traditions 1949

DeGivry, G., Witchcraft, Magic & Alchemy  Dover 1971

Diggs, E., Black Chronology  Hall 1983

Diop, C.A., The African Origin of Civilization: Myth Or Reality  Lawrence Hill Books 1974

Du Bourguet, P., Early Christian Painting  Viking 1965

Dunand, F., & Lichtenberg, R., Mummies: A Voyage Through Eternity  Thames & Hudson 1994

Eban, A., Heritage: Civilization and the Jews  Summit Books 1984

Edwards, I., The Pyramids of Egypt  Viking 1947, 1993

El Mahdy, C., Mummies, Myth and Magic  Thames & Hudson 1993

Erman, A., A Handbook of Egyptian Religion  Longwood Press 1977

Estes, J., The Medical Skills of Ancient Egypt  Science History Publ. 1989

Evans, J., Life in Medieval France  Phaidon 1969

Fairservis, W., Mesopotamia: The Civilization that Rose Out of Clay  Macmillan 1964

Fairservis, W., The Ancient Kingdoms of the Nile: And the Doomed Monuments of Nubia  Crowell Publ. 1962

Ferguson, E., Encyclopedia of Early Christianity  Garland 1990

Fernie, W., The Occult and Curative Power of Precious Stones  Harper & Row 1973

Finch, C. Ancient African Medical Practices: Lecture (audio) 1992

Fischer, H., An Elusive Shape within the Fisted Hands of Egyptian Statues  Metropolitian Museum Journal 1975

Frank, T., A History of Rome  Holt 1923

Fraser, P., Eratosthenes of Cyrene  Oxford Univ. 1970

Gardiner, A., Egypt of the Pharaohs  Oxford Press 1971

Ghalioungui, P., Magic and Medical Science in Ancient Egypt  Barnes & Noble 1963

Ghazanfar, A., Primate Audition: Ethnology and Neurobiology  CRC 2002

Goldman, J., Holy Harmony (CD)  Goldman & Sound Healers Publishing 2002

Goodman, F., Magic Symbols  Trodd Publ. House 1989

Gordon, C., Before the Bible  Harper & Row 1962

# Bibliography

Gouldner, A., The Hellenic World  Basic Books 1965

Grant, M., The History of Ancient Israel  Scribner & Sons 1984

Grant, M., The Roman Emperors  Scribner & Sons 1985

Gray, J., Near Eastern Mythology  Bedrick 1985

Grossinger, R., Planet Medicine: From Stone Age Shamanism to Post-Industrial Healing  Anchor Books 1980

Guillamont, A., The Gospel According to Thomas  Harper & Bros. 1959, 1984

Gunn, B., The Instruction of Ptahhotep and the Instruction of Kegemni: The Oldest Books in the World  Murray 1908

Gurney, J., Kingdoms of Asia the Middle East and Africa  Crown Publ. 1986

Haley, A., Roots  Dell 1974

Hall, M., The Secret Teachings of All Ages  The Philosophical Research Soc. 1977

Harnack, A., The Mission and Expansion of Christianity in the first three centuries Vol. II  Williams & Norgate 1908

Harris, G., Ancient Egypt  Facts on File 1990

Hastings, J., Selbie, J., & Lambert, J., Dictionary of the Apostolic Church  Scribner 1916

Hentz, F., & Long, G., Experiments with Chemical Reactions  Paladin House 1985

Heymsfield, S., Lohman, T., Wang, Z., Going, S., Human Body Composition  Human Kinetics 2005

Hicks, J., The Persians  Timelife 1975

Higgins, G., Anacalypsis  University Books 1965

Hinz, W., The Lost World of Elam  N.Y. Press 1973

Hobson, C., The World of the Pharaohs  Thames & Hudson 1982

Hodgson, M., The Venture of Islam: Conscience and History in a World Civilization  U. of Chicago 1974

Horowitz, L., DNA: Pirates of the Sacred Spiral (DVD)  New Science Ideas 2005

Horowitz, L., & Puleo, J., Healing Codes for the Biological Apocalypse  Tetrahedron 2006

Hurtak, J., The books of knowledge: the keys of Enoch  Academy for Future Science 1977

Hyde, W., Paganism to Christianity in the Roman Empire  Univ. of Penn. 1946

Iakovids, S., Mycenae - Epidaurus  Ekdotike 1978

Ions, V., Egyptian Mythology  Bedrick Books 1982

Irram, S., Death and Burial in Ancient Egypt  Longman 2003

Isocrates, Busiris  Norlin, G., (tr) http://classics.mit.edu/Isocrates/isoc.11.html 2-3-99

Jackson, J., An Introduction to African Civilizations  University Books 1970

Jobes, G., Dictionary of Mythology, Folklore and Symbols  Scarecrow Publ. 1961

Johnson, R., Atomic Structure  Twenty-First Century 2007

Jordan, M., Encyclopedia of Gods: Over 2500 Deities of the World  Facts of File 1993

Kaster, J., Wings of the Falcon: Life and Thought of Ancient Egypt  Holt, Rinehart & Winston 1968

Kedourie, E., Nationalism in Asia and Africa  Routledge 1974

# Bibliography

Kephart, C., Races of Mankind  N.Y. Philosophical Lib. 1960

Kidd, B., Documents Illustrative of the History of the Church Vol. II  Soc. for Promoting Christian Knowledge 1922

King, A., Quotations in Black  Greenwood Press 1981

Kramer, S., From the Tablets of Sumer  Falcon Wing Press 1956

Kristeller, P., Renaissance Thought and Its Sources  Columbia 1979

Lacroix, P., Science and Literature in the Middle Ages and at the Period of the Renaissance  London 1878

Lakhovsky, G., & Clement, M., Secret of Life: Cosmic Rays and Radiations of Living Beings  William Heinemann 1939

Lakhovsky, G., & Clement, M., Secret of Life: Cosmic Rays and Radiations of Living Beings  Mokelumme Hill 1970

Lucas, A., Ancient Egyptian Materials & Industry  E. Arnold Co. 1926

Lynes, B., Cancer Solutions: Rife, Energy Medicine and Medical Politics  Elsmere Press 2000

Lynes, B., The Cancer Cure That Worked: Discovery and Suppression of the Cancer Cure That Worked!  Marcus Books 2005

MacCulloch, C., The Mythology of All Races  Cooper Square Publ. 1964

Malcioln, J., The African Origins of Modern Judaism  Africa World Press 1996

Malek, J., & Forman, W., In the Shadow of the Pyramids  Univ. of Oklahoma 1986

Maspero, G., History of Egypt  The Groliers Soc. 1901
Maxwell, J., That Old Time Religion  Book Tree 2000
Mc Coy, R., Under the Tamarisk Tree  Stout State U. 1970
McGrew, R., Encyclopedia of Medical History  McGraw Hill 1985
Mertz, B., Red Land, Black Land  McCann 1966
Mertz, B., Temples, Tombs and Hieroglyphs  Mertz 1990
Meyer, M., The Gnostic Gospels of Jesus  Meyer 2005
Mialon, E., The Great Pharaoh Ramses and His Time  Exim Publ. 1985
Mieroop, M., King Hammurabi of Babylon  Blackwell 2005
Millar, F., The Emperor in the Roman World  Cornell Univ. Press 1977
Montet, P., Eternal Egypt  The New Amer. Lib. 1964
Montet, P., Lives of the Pharaohs  World Publ. 1968
Morelle, R., Animal world's communication kings  BBC http://news.bbc.co.uk/1/hi/sci/tech/3430481.stm 2007
Moscati, S., The Face of the Ancient Orient  Anchor Books 1962
Mozley, C., Tales of Ancient Egypt  Watts Publ. 1960
Muck, O., The Secret of Atlantis  Time Books 1976
Muller, M., Mythology of All Races Vol. XII  Cooper Square 1964
Murphy, E., Diodorus on Egypt  McFarland 1985
Nardo, D., Atoms  Kidhaven Press 2002
Noblecourt, C., Tutankhamen: Life and Death of a Pharaoh  Penguin 1963
Oates, J., Babylon  Thames & Hudson 1979

# Bibliography

Ogden, J.,  Jewellery of the Ancient World  Rizzoli 1982

Perl, L.,  Mummies, Tomb and Treasure  Clarendon 1987

Petrie, F.,  Wisdom of the Egyptians  British School of Arch. 1940

Pfeffer, L.,  Church, State, and Freedom  Beacon Press 1953

Prentice, W.,  Therapeutic Modalities in Rehabilitation  McGraw-Hill 2005

Quirke, S.,  Ancient Egyptian Religion  Dover 1992

Quirke, S.,  Who were the Pharaohs  Dover 1990

Randolph, P.,  Hermes Mercurius Trigismestus: his Divine Pymander  Randolph 1871 - 1889

Rapacholi, M.,  Essentials of Medical Ultrasound: A Practical Introduction to the Principles, Techniques and Biomedical Applications  Humana Press 1982

Rashdall, H.,  The Universities of Europe in the Middle Ages  Oxford Univ. 1936

Rawlinson, G.,  History of Ancient Egypt  Dodd & Mead 1862

Rawlingson, G.,  History of Herodotus: A New English Version  Murry 1880

Reeves, N.,  Into the Mummy's Tomb  Madison Press 1992

Richardson, M.,  Hammurabi's Laws: Text, Translation and Glossary  C. I. P. Group  2000

Rogers, J.A.,  Sex and Race  Rogers Publ. 1967

Rosten, L.,  Joys of Yiddish  Pocket Books 1970

Rumsey, F.,  & McCormick, T.,  Sound and Recording: An Introduction  Elsevier 2006

Sadoul, J.,  Alchemists and Gold  Putnam 1972

Saggs, H., Civilization Before Greece and Rome  Yale Univ. 1989

Saunders, J., The Transitions From Ancient Egyptian to Greek Medicine  Univ. of Kansas 1963

Sauneron, S., Priests of Ancient Egypt  Grove Press 1960

Sayce, A., The Religions of Ancient Egypt and Babylonia  Clark 1903

Schulberg, L., Historic India  Time 1968

Sedgwick, W., Tyler, H., A Short History of Science  Macmillian 1917

Sedgwick, W., Tyler, H., & Bigelow, R., A Short History of Science  Macmillian 1939

Seldes, G., The Great Quotations  Pocket Books 1960, 1967

Sewell, B., Egypt Under the Pharaohs  Putnam 1969

Singer, C., A History of Technology  Clarendon Press 1954

Sjoo, M., & Mor, B., The Great Cosmic Mother  Harper & Row 1817

Snowden, F., Blacks in Antiquity  Harvard Univ. Press 1970

Spence, L., The Occult Sciences in Atlantis  Weiser Inc. 1970

Stanfield, R., Gorny, M., Pazner, S., & Wilson, I., Crystal Structures of Human Immunodeficiency Virus Type 1 (HIV-1) Neutralizing Antibody 2219 in Complex with Three Different V3 Peptides Reveal a New Binding Mode for HIV-1 Cross-Reactivity  American Society for Microbiology Vol. 80(12) Jun. 2006

Stein, W., Atlantis  Greenhaven Press 1989

Stemman, R., Atlantis and Lost Lands  Doubleday 1977

# Bibliography

Sykes, E., Everyman's Dictionary of Non-classical Mythology Dutton & Sons 1952

Tesla, N., & Childress, D., The Fantastic Inventions of Nikola Tesla Adventures Unlimited 1993

Thompson, W., Medicines From the Earth: A Guide to Healing Plants Harper & Row 1983

Thornwald, J., Science and Secrets of Early Medicine Harcourt, Brace & World 1963

Tompkins, P., The Magic of Obelisks Harper & Row 1981

Tompkins, P., The Secret of the Great Pyramid Harper & Colophon Books 1978

Tong, L., Pav, S., Pargellis, C., Lamarre, F., & Anderson, P., Crystal structure of human immunodeficiency virus (HIV) type 2 protease in complex with a reduced amide inhibitor and comparison with HIV-1 protease structures. Proceedings of the National Academy of Sciences Vol. 90(18) Sep. 15, 1993

Trigger, B., Kemp, B., O' Connor, D., & Lloyd, A., Ancient Egypt: A Social History Cambridge Univ. 1983

Uvarov, V., Wands of Horus Uvarov 2005

Van Sertima, I., Blacks in Science: Ancient and Modern Transactions 1983

Veith, W., The Man Behind the Mask (DVD) Amazing Discoveries 2004

Waddell, W., Manetho Harvard Univ. Press 1980

Waddell, W., Manetho, Ptolemy Harvard Univ. Press 1948

Walker, B., The Woman's Encyclopedia of Myths and Secrets Rodeo Press 1981

Ward, P., A Dictionary of Common Fallacies  Prometheus Books 1989

Watson, T., Therapeutic Ultrasound http://www.electrotherapy.org/downloads/Modalities/Therapeutic%20Ultrasound.pdf 2006

Whiston, W., The Life and Works of Flavius Josephus  Holt, Rinehart & Winston 1957

Wieland, C., Has an ape learned to talk? Creation Ministries International  http://www.creationontheweb.com/content/view/2750/ 2003

Wilson, J., Culture of Ancient Egypt  Phoenix Books 1951

Windsor, R., From Babylon to Timbuktu  Exposition Press 1969

Woolsey, J., Symbolic Mythology  Woolsey 1917

Worthy, R.L., About Black Hair  KornerStone Books 2006

Wright, G., Biblical Archaeology  Redwood Burn 1974

Zahoor, A., Hospitals and Medical Schools in the Dark and Middle Ages  http://users.erols.com/gmqm/sibai10.html 8-22-99

Zemlin, W., Speech and Hearing Science: Anatomy and Physiology  Prentice Hall 1988

_____., An Apparent New Direct Human Ancestor Fossil  Science Week  Vol. 5, No. 28, July 13, 2001  WWW.scienceweek.com/2001/sw010713.htm  2007

_____., Anchor Bible Dictionary  Doubleday 1992

_____., Ancient Egypt Fan: Ancient Egyptian Medical Papyri  http://indigo.ie/~marrya/papyri.html  2007

# Bibliography

_____., Ancient Lives (VHS) Spry-Leverton & WTTW 1984, 1988

_____., Ancient Mesopotamia  Nauka Publ. 1969

_____., Arthur Clarke's Mysterious World: Ancient Discoveries (VHS) Discovery Chan. 9/21/87

_____., Atlas of the Human Body  Barron's 2006

_____., Behind the Name: the etymology and history of first names  www.behindthename.com/name/hammurabi 2007

_____., Cambridge Dictionary of Philosophy  Cambridge Univ. 1959

_____., Caxton's History of the World  Vol. I  New Caxton L.T.D. 1969

_____., Coil For Electromagnets  Twenty First Century Books www.tfcbooks.com/patents/coil.htm  2007

_____., Collier's Encyclopedia  Macmillan 1986

_____., Compton's Encyclopedia  Compton's Learning Co. 1991

_____., Dictionary of Scientific Biography  Scribner & Sons 1970, 1976

_____., Encyclopaedia Britannica  Online  Larynx www.library.eb.com/article-9047232 2007

_____., Encyclopaedia Judaica  Macmillan 1971

_____., Encyclopedia Britannica Encyclopaedia Britannica Inc. 2007

_____., Encyclopedia of Biodiversity  Vol. I,  Academic Press 2001

𝔜𝔈𝔖𝔘𝔈

_____., Encyclopedia of Medical History  McGraw Hill 1985

_____., Encyclopedia of Philosophy  Macmillan Freepress 1973

_____., Encyclopedia of Religion  Macmillan 1987

_____., Essential Atlas of Physiology  Barron's 2005

_____., Fativa: Ancient Healing Codes Revealed in Bible Are Published by Tetrahedron Press http:://Factiva.com 1999

_____., Great Soviet Encyclopedia  Macmillan 1985

_____., HarperCollins Encyclopedia of Catholicism 1995

_____., Hermes Mercurius Trismegistus: his Divine Pymander in Seventeen Books  T. Brewster 1657

_____., Historical Atlas of Africa  Cambridge Univ. 1985

_____., Illustrated Dictionary & Concordance of the Bible  Jerusalem Publ. House 1986

_____., Jordan Maxwell:  Egypt in the New Millennium Ancient Wisdom  (DVD) 1999

_____., Loyola Medicine:  Kidney Stones http://www.luhs.org/depts/urology/kidney_stones.htm 2007

_____., Monkeys 'grasp basic grammar' Monkeys can understand the simple rules of grammar but the key element of all human languages is beyond them, a study at Harvard University has shown.  BBC Science/News http://news.bbc.co.uk/1/hi/sci/tech/3413865.stm  2004

_____., Mystic Places  Time Life Books 1987

_____, Naked Truth  (VHS) I.R.E.S. 1991

# Bibliography

_____., NASA http://www.grc.nasa.gov/WWW/k-12/airplane/sound.html 2007

_____., New Catholic Encyclopedia  Catholic Univ. of America 2003

_____., NIAAA  http://www.niaaa.nih.gov/ 2007

_____., Nova:  Pyramids: Interview with Dr. Zahi Hawass, Director of the Pyramids  http://www.pbs.org/wbgh/nova/pyramid/excavation/hawass.html 2007

_____., Physics Classroom tutorial: The Nature of a Wave  http://www.glenbrook.k12.il.us/GBSSCI/PHYS/Class/waves/u10l1b.html#energy 2007

_____., Physlink.com:  What is the physics involved with breaking glass with your voice?  http://www.physlink.com/ 2007

_____., Roots  (VHS)  Wopler Productions 1981

_____., Shocking Treatment Proposed for Aids.  (Zapping the Aids Virus with low voltage electric current).  Science News  Vol. 139, No. 13 Mar. 30, 1991  p. 207

_____., The American Peoples Encyclopedia  Spencer Press 1955

_____., The Beck Protocol  Sharing Health (DVD) 2001

_____., The Black Virgin/Black Madonna  http://shell.rmi.net/~ma3/index.html 1-20-99

_____., The Cambridge Ancient History Volume I Part II: Early History of the Middle East  Cambridge Univ. 1989

_____., The Columbia Encyclopedia  Columbia Univ. Press 1945

_____., The Encyclopedia of Earth Biodiversity www.eoearth.org/article/Biodiversity 2007

_____., The Granada Forum: Suppressed Medical Discovery (DVD) Transformation Technologies 1997

_____., The Jewish Encyclopedia  Funk & Wagnalls 1903

_____., The Kybalion: Hermetic Philosophy  Yoga 1912

_____., The McGraw-Hill Encyclopedia of World Biography  McGraw-Hill 1973

_____., The New Columbia Encyclopedia  Columbia Press 1975

_____., The Riddle of the Dead Sea Scrolls  Discovery Chan.  Mitchell 1990

_____., The Strecker Memorandum  (VHS)  The Strecher Group 1988

_____., The World Book Encyclopedia  The World Book Encyclopedia 1959

_____, Transformation of Myth Through Time: From the Id to the Igo in the Orient  (VHS) W. Free 1989

_____., Universal Standard Encyclopedia  W. Funk 1955

_____., Webster's Biographical Dictionary  Merriam Co. 1980

_____., Webster's Medical Dictionary  Wiley Publ. 2003

_____., Wikipedia: Larynx http://en.wikipedia.org/wiki/larynx 2007

_____., Wikipedia: Throat http://en.wikipedia.org/wiki/Throat 2007

_____., Wikipedia: Vocal Folds http://en.wikipedia.org/wiki/Vocal_cords 2007

# Index

# INDEX

## A

| | |
|---|---|
| Abraham | 84 |
| Adam | 14 |
| Aids | 39 |
| Akbar, N. | 197, 198, 199 |
| Akhenaten | 17, 18 |
| Akkadian | 75, 76, 80, 169 |
| Akkadian Words | 74 |
| Alphabet | 80 |
| Amen | 149 |
| Amorites | 168 |
| Amplification | 28, 31, 40, 93, 94 |
| Anba Hidra | 149 |
| Ankh | 198 |
| Anthropoids | 6 |
| Aor | 140 |
| Arabic | 76 |
| Aramaic | 75, 78, 80 |
| Aramaic Words | 74 |
| Arameans | 170 |
| Archimedes | 180 |
| Aria | 83 |
| Aristotle | 180 |
| Arnald of Villanova | 189 |
| Asclepius (*see* Imhotep) | |
| Assyria | 167 |
| Atom | 101, 102, 175 |
| Atum | 174, 175 |

## B

| | |
|---|---|
| Ba | 119, 120, 121, 122, 123 |
| Babel | 71, 72 |
| Baghdad Battery | 135 |
| Battery | 131, 134, 135, 136 |
| Beck Protocols | 44, 125, 126 |
| Bernald de Gordon | 189 |
| Black hair | 108, 109, 110, 111, 194, 195, 196 |

224

# Index

Blood Electrification 43, 44, 112, 121, 123 - 127, 136, 140, 144
Book of II Corinthians 88
Book of Genesis 13, 14, 49
Book of John 13, 48, 49
Book of Luke 73, 191
Book of Mark 89
Book of Matthew 89
Book of Psalms 15
Book of Romans 69
Boyle, R. 131
Brown, L. 46
Buddhists 194

Chemistry 115, 116, 123, 173, 174
Chimpanzee 6, 7, 8
Chinese 23, 73
Christ 33, 86, 160
Constantine 160, 161
Coptics 149, 151
Cuneiform 2, 3, 169
Cush (see Kush)

## C

Canaanite 75, 78, 82, 170, 175, 195
Cancer 93, 98, 99
Carthage 82
Cartouche 21
Chakra 192

## D

Darius I 83
Democritus 175, 180
Dies Solis 161
Diocletian 142
Diodorus 175, 179, 181, 186

## INDEX

DNA 6
38, 101, 102, 103, 104, 105, 108, 109, 112, 194

### E

Eardrum 27
Earth 50
Easter 162
Edomites 170
Eggebrecht, A. 135
Elam 83
84, 167, 168, 169
Electromagnetic Spectrum 98
Electroplating 134
Energy Wave Device 126
Essenes 70
ESWL 35
36, 40
Ethiopia 80
83
Ethiopian Words 74
Euclid 178
Eudoxos 176
180
Euphrates 168

### F

Followers of the Way 70
157
Foord, A. 99
Franklin, B. 132
Frequency 30
31, 38, 40, 41, 66, 93, 94, 96, 97, 98, 105, 107, 115

### G

Glass 137
138
Gorillas 6
Gray, S. 131
Gregorian Calendar 162
163
Gregory XIII 162

### H

Hallelujah 149

# Index

Hammurabi 19 20, 169
Harris, J. 139
Hathor 200
Hathor (Temple of) 139 140
Hatsheput 16 17
Hawass, Z. 142
Hebrew Words 74 149
Hecataeus 180
Herodotus 122 178, 179
Hertz 30 31
Hieroglyphs 2 3, 21
Higgins, G. 86
Hippocrates 184 185
HIV v 39, 40, 41, 42, 43, 44
Homer 180
Hominoids 6
Horowitz, L. 101 102, 103, 104, 105, 108, 109 112 192, 194
Horus 114 120, 121, 122, 123, 130, 149
Hosea 156

## I

Iesous 55 56, 66
IHVH (*see* YHVH)
Imhotep 184 185
Indus 83
Ionians 179
Iran 167
Isaiah 85

## J

Jehovah 54 59, 66
Jericho 32
Jesus 55 56, 61, 62, 65, 66
Johnson, M. 99
Judas 42
Julian Calendar 162

## INDEX

### K

| | |
|---|---|
| Ka | 119 |
| 120, 121, 122, 123 | |
| Kaali, S. | 43 |
| 44 | |
| Khemet | 123 |
| Konig, W. | 134 |
| 135 | |
| Kunta Kinte | 23 |
| Kush | 80 |
| 83 | |

### L

| | |
|---|---|
| Lakhovsky, G. | 92 |
| 93, 94, 103, 112, 124 | |
| Larynx | 7 |
| 8 | |
| Lorenzen, L. | 105 |
| Lycurgos | 180 |
| Lyman, W. | 43 |

### M

| | |
|---|---|
| Madonna | 86 |
| 150 | |
| Media | 83 |
| 167 | |
| Meditation | 36 |
| 37 | |
| Melanin | 109 |
| 194 | |
| Menkaura | 116 |
| Metallurgy | 128 |
| 129, 137, 142, 174 | |
| Mi | 105 |
| Milky Way | 47 |
| 48 | |
| Minoans | 80 |
| Mochus | 175 |
| Monastery | 149 |
| Moors | 189 |
| 190 | |
| Montpellier | 189 |
| MOR | 98 |
| Moses of Khorene | 83 |
| Musaeus | 180 |

# Index

## N

| | |
|---|---|
| Nebuchadnezzar | 169 |
| Nefretari | 133 |
| Nodon, A. | 124 |
| Nubians | 80, 81 |

## O

| | |
|---|---|
| Obelisk | 141 |
| Oenopides | 175, 180 |
| Omoro | 23 |
| Oriental | 78 |
| Orpheus | 180 |

## P

| | |
|---|---|
| Pakistan | 83 |
| Paul | 88, 122 |
| Persia | 83, 167, 169 |
| Phoenicians | 78, 79, 82, 175 |
| Pitch | 28, 30, 31, 51 |
| Plato | 180 |
| Plutarch | 177 |
| Primates | 4, 9 |
| Pyramids | 80, 118, 141, 142, 176 |
| Pythagoras | 177, 178, 180 |

## Q

## R

| | |
|---|---|
| Radio Wave | 39, 40, 93, 97, 98 |
| Rameses II | 20, 21, 22, 127 |
| Resonance | 28, 29, 30, 31, 36, 97 |
| Rhind, Papyrus | 143 |
| Rife, R. | 40, 41, 96, 97, 98, 99, 100, 103, 112 |

RNA 108
Rods of Ancient Egypt
(see Wands of Horus)

## S

Sabbath 160
161, 162
Sacred Languages 73
75
Sanskrit 73
Schwolsky, P. 44
Seti I 196
200
Sidon 175
Sinyukhin, A. 95
103, 112
Socrates 177
Solon 180
Spirit 119
Stones 36
Strabo 143
175
Strecker, R. 40
41, 42, 43
Susiana (Susa) 83
167, 168, 169

## T

Tesla, N. 106
107, 108, 141, 142
Tetragrammaton 53
54
Thebes 181
Thomas 157
162
Thoth 192
Thymus 38
Tibetan 73
Tigris 83
167, 168
Tomb Lighting 132
133, 139, 140
Tone 28
31, 51, 65, 66
Trachea 8
Tutankhamen 129
137

## U

Ultrasound 36

# Index

Ur 168
Uraeus 196

## V

Vacuum 138
Valley of the Kings 133
Vocal Chords 7 8, 33
Volta, A. 132, 136
Voltaic Cell Battery 136

## W

Wands of Horus 113 114, 115, 116, 117, 118, 120, 121, 122, 123, 124, 125, 128, 130, 131, 140
World Health Org. 42

## X

## Y

Yahweh 54 58, 66
Yashua (Yeshua) 60 66
YHSVH 42 52, 55, 56, 59, 61, 65, 66, 70, 73, 83, 86, 87, 88, 89, 148, 150, 156, 157
YHVH 12 37, 38, 52, 54, 55, 58, 66, 67, 70, 71, 73, 89, 148, 156, 157, 161

## Z

Zeus 56
Zodiac 193

# Past Reviews of the Author's Work!

*It is Great!*
**Jenkins Group Reviewer**

*Mr. Worthy's research is both—scholarly and important!*
**Ms. Pearl Randall, Former Editorial Coordinator**
**Addison-Wesley**

*Beloved R.L. Worthy – Thank you for sharing Truth!*
**Iyanla Vanzant's Inner Visions Team Member**

*I love the Book!*
**Alicia Banks – Radio Talk Show Host**

*Thank you for the wonderful Book!*
**Vivian Stringer – Rutgers Women's Basketball Coach**

*This is a very interesting book!*
**Megan Sukys – Radio Talk Show Host**

*A lot has been written on the subject but you're shedding new light!*
**Barbara Thomas - Northwest African Amer. Museum**

*Mr. Worthy's ongoing commitment to share much wisdom from experience as another of our vindicationist scholars engage the primary representation of our peoplehood and forge to share our destiny through telling our story—*
**Ndugu G.B. T'Ofori-Atta, Professor**
**Interdenominational Theological Center**

# YHSVH

## Discount Order Form

Of course, you can call any of your local bookstores and ask them to order a copy of *YHSVH* for you for 29.95 plus tax. You should be able to pick your copy up in 48-hours. However, if you would like to have a copy (or copies) of *YHSVH* delivered to your doorstep and save money—feel free to photocopy this page and use it as an order form. Just fill in the information and mail the <u>discount sale price</u> of $21.95 plus $3.00 shipping and handling for each book to the address below (<u>money orders please</u>). You should receive your order in about a week. For orders of 5 books or more - contact us via email to learn about our volume discounts and <u>online credit card ordering</u> process.

Name:_____

Address:_____

City, State & Zip:_____

Number of Books:_____    Amount Enclosed:_____

*{No sales tax outside of Wa. State}*

**KornerStone Books**
**6947 Coal Creek Pkwy.**
**Suite 206**
**Newcastle, WA. 98059**
**Ksbooks@execs.com**

# *About* **Black Hair**

## Discount Order Form

Of course, you can call any of your local bookstores and ask them to order a copy of *About Black Hair* for you for 18.95 plus tax. You should be able to pick your copy up in 48-hours. However, if you would like to have a copy (or copies) of *About Black Hair* delivered to your doorstep and save money—feel free to photocopy this page and use it as an order form. Just fill in the information and mail the <u>discount sale price</u> of $12.95 plus $3.00 shipping and handling for each book to the address below (<u>money orders please</u>). You should receive your order in about a week. For orders of 5 books or more - contact us via email to learn about our volume discount and <u>online credit card ordering</u> process.

Name:_____

Address:_____

City, State & Zip:_____

Number of Books:_____    Amount Enclosed:_____

*{No sales tax outside of Wa. State}*

**KornerStone Books
6947 Coal Creek Pkwy.
Suite 206
Newcastle, WA. 98059
Ksbooks@execs.com**

# The Founders' Facade

## Discount Order Form

Of course, you can call any of your local bookstores and ask them to order a copy of *The Founders' Façade* for you for 21.95 plus tax. You should be able to pick your copy up in 48-hours. However, if you would like to have a copy (or copies) of *The Founders' Facade* delivered to your doorstep and save money—feel free to photocopy this page and use it as an order form. Just fill in the information and mail the <u>discount sale price</u> of $14.95 plus $3.00 shipping and handling for each book to the address below (<u>money orders please</u>). You should receive your order in about a week. For orders of 5 books or more - contact us via email to learn about our volume discount and <u>online credit card ordering</u> process.

Name:_____

Address:_____

City, State & Zip:_____

Number of Books:_____    Amount Enclosed:_____

*{No sales tax outside of Wa. State}*

**KornerStone Books**
**6947 Coal Creek Pkwy.**
**Suite 206**
**Newcastle, WA. 98059**
*Ksbooks@execs.com*

www.ingramcontent.com/pod-product-compliance
Lightning Source LLC
Chambersburg PA
CBHW070940230426
43666CB00011B/2505